Sustainable ICT Adoption and Integration for Socio-Economic Development

Charles K. Ayo
Covenant University, Nigeria

Victor Mbarika
Southern University, A & M College, USA

A volume in the Advances in Electronic Government, Digital Divide, and Regional Development (AEGDDRD) Book Series

IGI GLOBAL
DISSEMINATOR OF KNOWLEDGE
www.igi-global.com

Published in the United States of America by
IGI Global
Information Science Reference (an imprint of IGI Global)
701 E. Chocolate Avenue
Hershey PA, USA 17033
Tel: 717-533-8845
Fax: 717-533-8661
E-mail: cust@igi-global.com
Web site: http://www.igi-global.com

Library of Congress Cataloging-in-Publication Data

Names: Ayo, Charles K., 1958- editor. | Mbarika, Victor W. A. (Victor Wacham A.), editor.
Title: Sustainable ICT adoption and integration for socio-economic development / Charles K. Ayo and Victor Mbarika, editors.
Description: Hershey, PA : Information Science Reference, [2017]
Identifiers: LCCN 2017005921| ISBN 9781522525653 (hardcover) | ISBN 9781522525660 (ebook)
Subjects: LCSH: Internet in public administration--Developing countries. | Internet in education--Developing countries. | Information technology--Economic aspects--Developing countries. | Information technology--Social aspects--Developing countries.
Classification: LCC JF1525.A8 S92 2017 | DDC 303.48/33091724--dc23 LC record available at https://lccn.loc.gov/2017005921

This book is published in the IGI Global book series Advances in Electronic Government, Digital Divide, and Regional Development (AEGDDRD) (ISSN: 2326-9103; eISSN: 2326-9111)

British Cataloguing in Publication Data
A Cataloguing in Publication record for this book is available from the British Library.

Advances in Electronic Government, Digital Divide, and Regional Development (AEGDDRD) Book Series

ISSN:2326-9103
EISSN:2326-9111

Editor-in-Chief: Zaigham Mahmood, University of Derby, UK & North West University, South Africa

MISSION

The successful use of digital technologies (including social media and mobile technologies) to provide public services and foster economic development has become an objective for governments around the world. The development towards electronic government (or e-government) not only affects the efficiency and effectiveness of public services, but also has the potential to transform the nature of government interactions with its citizens. Current research and practice on the adoption of electronic/digital government and the implementation in organizations around the world aims to emphasize the extensiveness of this growing field.

The Advances in Electronic Government, Digital Divide & Regional Development (AEGDDRD) book series aims to publish authored, edited and case books encompassing the current and innovative research and practice discussing all aspects of electronic government development, implementation and adoption as well the effective use of the emerging technologies (including social media and mobile technologies) for a more effective electronic governance (or e-governance).

COVERAGE

- E-Government in Developing Countries and Technology Adoption
- Knowledge Divide, Digital Divide
- Public Information Management, Regional Planning, Rural Development
- Issues and Challenges in E-Government Adoption
- Current Research and Emerging Trends in E-Government Development
- Emerging Technologies within the Public Sector
- Adoption of Innovation with Respect to E-Government
- Frameworks and Methodologies for E-Government Development
- Case Studies and Practical Approaches to E-Government and E-Governance
- ICT within Government and Public Sectors

IGI Global is currently accepting manuscripts for publication within this series. To submit a proposal for a volume in this series, please contact our Acquisition Editors at Acquisitions@igi-global.com or visit: http://www.igi-global.com/publish/.

The Advances in Electronic Government, Digital Divide, and Regional Development (AEGDDRD) Book Series (ISSN 2326-9103) is published by IGI Global, 701 E. Chocolate Avenue, Hershey, PA 17033-1240, USA, www.igi-global.com. This series is composed of titles available for purchase individually; each title is edited to be contextually exclusive from any other title within the series. For pricing and ordering information please visit http://www.igi-global.com/book-series/advances-electronic-government-digital-divide/37153. Postmaster: Send all address changes to above address.

Titles in this Series

For a list of additional titles in this series, please visit:
http://www.igi-global.com/book-series/advances-electronic-government-digital-divide/37153

Digital Governance and E-Government Principles Applied to Public Procurement
Rajesh Kumar Shakya (The World Bank, USA)
Information Science Reference • ©2017 • 355pp • H/C (ISBN: 9781522522034) • US $185.00

Politics, Protest, and Empowerment in Digital Spaces
Yasmin Ibrahim (Queen Mary, University of London, UK)
Information Science Reference • ©2017 • 363pp • H/C (ISBN: 9781522518624) • US $180.00

Handbook of Research on Sub-National Governance and Development
Eris Schoburgh (University of the West Indies, Jamaica) and Roberta Ryan (University of Technology Sydney, Australia)
Information Science Reference • ©2017 • 625pp • H/C (ISBN: 9781522516453) • US $285.00

Agent-Based Modeling in Humanitarian Interventions Emerging Research...
John McCaskill (University of Texas at Dallas, USA)
Information Science Reference • ©2017 • 99pp • H/C (ISBN: 9781522517825) • US $115.00

Impacts of the Media on African Socio-Economic Development
Okorie Nelson (Covenant University, Nigeria) Babatunde Raphael Ojebuyi (University of Ibadan, Nigeria) and Abiodun Salawu (North-West University, South Africa)
Information Science Reference • ©2017 • 328pp • H/C (ISBN: 9781522518594) • US $175.00

Population Growth and Rapid Urbanization in the Developing World
Umar G. Benna (Ahmadu Bello University, Nigeria) and Shaibu Bala Garba (Qatar University, Qatar)
Information Science Reference • ©2016 • 358pp • H/C (ISBN: 9781522501879) • US $205.00

For an enitre list of titles in this series, please visit:
http://www.igi-global.com/book-series/advances-electronic-government-digital-divide/37153

IGI GLOBAL
DISSEMINATOR OF KNOWLEDGE

www.igi-global.com

701 East Chocolate Avenue, Hershey, PA 17033, USA
Tel: 717-533-8845 x100 • Fax: 717-533-8661
E-Mail: cust@igi-global.com • www.igi-global.com

List of Reviewers

Marion Adebiyi, *Covenant University, Nigeria*
Ibukun Afolabi, *Covenant University, Nigeria*
Mayowa Agboola, *Covenant University, Nigeria*
Mosun Akinbode, *Covenant University, Nigeria*
Boluwaji Akinnuwesi, *Lagos State University, Nigeria*
Ambrose Azeta, *Covenant University, Nigeria*
Onyeka Emebo, *Covenant University, Nigeria*
Ibukun Eweoya, *Covenant University, Nigeria*
Daniel Gberevbie, *Covenant University, Nigeria*
Efosa Idemudia, *Arkansas Tech University, USA*
Salah Kabanda, *University of Cape Town, South Africa*
Mayoka Kituyi, *Makerere University, Uganda*
Chinoye Moses, *Covenant University, Nigeria*
Babafemi Odusote, *Covenant University, Nigeria*
Mercy Ogbari, *Covenant University, Nigeria*
Adeyemi Ogundipe, *Covenant University, Nigeria*
Kemi Ogunsola, *University of Ibadan, Nigeria*
Sena Okuboyejo, *Covenant University, Nigeria*
Funke Oladipupo, *Covenant University, Nigeria*
Funminiyi Olajide, *University of Portsmouth, UK*
Aderonke Oni, *Covenant University, Nigeria*
Samuel Oni, *Covenant University, Nigeria*
Emike Oshionebo, *University of Lagos – Akoka, Nigeria*
Femi Oyelami, *Bowen University, Nigeria*
Victoria Samuel, *Covenant University, Nigeria*

Table of Contents

Section 2
Using Information Technology to Support Secondary and Tertiary Education

Section 3
Contemporary ICT Application for Development

Detailed Table of Contents

Electronic government (e-government), facilitated through government websites are becoming the fastest delivery modes of government services, as they make services available to users, anywhere, anytime. Previous studies focused on the accessibility or quality of these websites in terms of web readiness and/or web presence quality measures. However, no known study has investigated the interrelationships among web readiness, web presence quality and information communication technology deployment for government services (ICT deployment) in Nigeria. The study used a questionnaire (test-retest Spearman r 0.79 for all construct items) to collect data from a sample of 117 public servants in 20 government agencies, on their knowledge of ICT deployment in their agencies; and a checklist for the content analysis of 207 Nigerian government websites, measuring their web readiness and web presence quality. The study recommended that government agencies should re-align ICT deployment with the information, services and features of their e-government websites.

This study examined the application of e-governance for service delivery in Ojo and Alimosho Local government areas of Lagos State. Questionnaire and Interviews were used as major instruments for data collection alongside other secondary sources. The simple Random Sampling technique was used to derive a sample of 100 respondents from a population size of 748 workers from Ojo and Alimosho Local Government Areas. Simple percentages were used for the demographic data analysis, as well as to test the hypotheses. The findings revealed that the application of E-governance is still a far cry in the local governments. Generally, the concept is novel among the workers as it is generally construed to mean electronic payment of salaries and lodgments into the Banks by the citizens. Beyond this, it was revealed that such facilities like web sites, constant electricity supply, enabling environments to boost tele-density, internet diffusion, adult computer literacy, cyber cafes, computer, etc. were not adequately available as expressed by the respondents.

Barriers such as unequal access, lack of digital skill, low income and disability constitute limiting factors for technology-mediated citizens and government interaction in developing countries. It is against this backdrop that this work explores the integration of Voice, Web and SMS technologies, in Nigeria's democratic process. The proposed system takes advantage of the ubiquitous nature of mobile devices to explore the plausibility of increasing the level of citizens' participation in democratic practices, particularly, those in rural areas with no Internet access and the physically challenged electorates. The server module for the e-democracy system was developed in PHP. Ozeki SMS server and Voxeo Voice server were used for SMS transaction code and VoiceXML code respectively. The prototype e-democracy system shows that developing nations can take advantage of their present level of technological development to give voice to the voiceless and improve their democratic system.

One of the major reasons for poor voter turn-up in transitional and post conflict countries in Africa is the credibility and legitimacy of the voter's register. Many voters are disfranchised, and many of the voter's registers are erroneous and lack credibility. This has created a wave of mistrust and discontentment amongst the voters and many prefer not to participate in the election process. Yet, the most fundamental form of exercising good democracy is through participating in regular free and fair elections. However, before a free and fair election is carried out, relevant electoral bodies and authorities normally establish a comprehensive, credible and inclusive voters' list, also called a voters' register. This chapter discusses manual voter registration in

Information and Computer Technology for Development (ICT4D), Human-Computer Interaction (HCI) and Critical Disability Studies are reviewed and, these theories are grounded in the authors' experience working with stakeholders in the region. The conclusion discusses promising future directions for participatory and collaborative research in Kenya, and more broadly in the East African context.

The purpose of this chapter is to share a successful case of how ICT was used for educational inclusiveness by students from previously disadvantaged backgrounds. This chapter further illuminates how students could be supported in acquiring digital and academic literacy skills, within a discipline context. Lecturers often find it difficult to define digital literacy beyond the use of ICT in a classroom, or understand how students' acquisition of digital literacy skills could be supported or further expanded into students' academic literacy skills. This piece of work therefore, provides insight to academics, researchers and students, on how: (1) ICT could be integrated into the curriculum; (2) digital literacy could be embedded into the curriculum; and (3) students could be supported in acquiring graduate attributes and skills that make them fit for the workplace.

This chapter explores Cloud computing adoption strategies for small and medium enterprises (SMEs) and microenterprise that can enhance the economic growth of local industries in developing countries. This focus is based on the recognitions that (1) economic growths of local industries reduce poverty through the creation of new jobs for poor people to earn a living wage and support their families, and (2) there is a lack of research on Cloud computing adoption strategy that addresses the unique constraints of small and medium enterprises (SMEs) in developing countries.

Prior studies have x-rayed the potentials and impetus for development that can be resultant from a full scale adoption of ICT in Africa particularly in Nigeria. Current challenges relating to infrastructures, cybercrime, government policies and so on that mitigate the benefits accruable from a virile ICT growth have also been highlighted. However, research also needs to really address possible areas of developmental benefit of ICT to rural target groups such as farmers, extension officers, health and social workers This paper explores possible alignment of ICT4D2.0 goals to ICT development in Nigeria. It posited that the digital divide that ICTD1.0 sets to bridge are still very evident at grassroots level. Adding computing and Internet functionality to technologies that already penetrate the grassroots such as mobiles phones, radios and televisions will enhance ICT4D2.0 goals in Nigeria. This will in turn increase productivity among this target group with attendant contribution to national development.

This study aimed at determining the level of computer skills acquired by secondary school students in Nsukka Local Government of Enugu State, Nigeria. Specifically the study sought to ascertain the level of word processing skills, graphic application skills and Internet skills acquired by secondary school students in Nsukka Local Government of Enugu State. Three research questions and three null hypotheses were formulated in line with the objectives to guide the study. The study employed a descriptive survey research design using a sample of 272 respondents drawn from three secondary schools selected for the study. The instrument used for data collection was a 41-item structured questionnaire with a reliability index of .72, .83, and .80 using split half method. Data collected was analyzed using mean and standard deviation for research questions while t-test statistic was used to test the null hypotheses at 0.05 level of significance. The study revealed that word processing skills and Internet skills are highly acquired by students' whereas graphic skills were moderately acquired by the students. It was therefore recommended that all

Preface

Research abounds on the nexus of ICTs and developmental issues, especially pertaining to developing nations. In the past decade, we have seen plethora of academic journals emerge with an agenda to publish work in this domain. Despite the widespread use of ICTs in developing nations, there still remains a debate on sustainability of these technologies. *Sustainable ICT Adoption and Integration for Socio-Economic Development* is intended to contribute to the sustainability discourse. The Editors, Prof. C. K Ayo and Prof. V. W. A. Mbarika, with over 300 combined peer-reviewed publications to their credit, bring together a team of esteemed scholars in domain ICTs for Development, to contribute and to bring to the readers exciting perspectives to the discourse.

This book reviews applications and case studies, of the use of information technology in Governance and Democracy, the use of Information Technology to Support Secondary and Tertiary Education and contemporary ICT Applications for Development. It highlights the importance of ICT in governance, economic development, and in education as well. This book is a blend of knowledge from different intelligent minds across the globe. From an obvious position, one would expect a book on information technology to be based on a whole lot of theories. However; experience seem to show that most logical and critical issues are best explained using examples and real life situations; which is the main purpose of this book. Diversity of minds and reasoning is a key strength of this book, since it helps one think out of the box and appreciates knowledge from different points of views. This book looks at different scenarios in which information technology comes in handy, especially in developing countries that still have great need for innovation. It examines the use and application of information technology across different fields of study.

This book provides profound knowledge and a comprehensive treatment. Chapter 1 looks at Assessment of Levels and Interrelationships of ICT Deployment, Web Readiness, and Quality eGovernment by Kemi Ogunsola, University of Ibadan, Nigeria. It explains Electronic government (e-government), facilitated through government websites are becoming the fastest delivery modes of government services, as they make services available to users, anywhere, anytime. Previous studies focused on the accessibility or quality of these websites in terms of web readiness and/or web presence quality measures. However, no known study has investigated the interrelationships among web readiness, web presence quality and information communication technology deployment for government services (ICT deployment) in Nigeria.

Chapters 2 looks at Interrogating the Application of E-Governance for Service Delivery by Francis F. Nchuchuwe, Lagos State University, Ojo, Nigeria. Chapter 3 looks at Development of an Inclusive Participatory Democratic System by Aderonke A. Oni, Charles K. Ayo, and Ambrose A. Azeta, all of Covenant University, Nigeria. Their research investigates barriers such as unequal access, lack of digital skill, low income and disability constitute limiting factors for technology-mediated citizens and government interaction in developing countries. They explore the integration of Voice, Web and SMS technologies, in Nigeria's democratic process. Their proposed system takes advantage of the ubiquitous nature of mobile devices to explore the plausibility of increasing the level of citizens' participation in democratic practices, particularly, those in rural areas with no Internet access and the physically challenged electorates. The evaluation of the e-democracy system shows that developing nations can take advantage of their present level of technological development to give voice to the voiceless and improve their democratic system.

In Chapter 4, Emmanuel Eilum of Makerere University, Uganda looks at Embracing Biometric Voter Registration in Africa: Pros and Cons. One of the major reasons for poor voter turn-up in transitional and post conflict countries in Africa is the credibility and legitimacy of the voter's register. Many voters are disfranchised, and many of the voter's registers are erroneous and lack credibility. This has created a wave of mistrust and discontentment amongst the voters and many prefer not to participate in the election process. Yet, the most fundamental form of exercising good democracy is through participating in regular free and fair elections. However, before a free and fair election is carried out, relevant electoral bodies and authorities normally establish a comprehensive, credible and inclusive voters' list, also called a voters' register. His work further discusses manual voter registration in Africa and its challenges. It further discusses biometric voter registration systems in different African countries; the opportunities it provides to pacify and redeem voters' registers. The author provides relevant advice to governments in Africa on the most critical success factors for adopting biometric voter registration.

Chapter 5 focuses on E-Learning Adoption Among Selected Secondary Schools by Adedeji O. Adewusi, Olabisi Onabanjo University, Nigeria. This chapter examines E-learning adoption in secondary schools of Ogun state. Constructivism Learning Theory was used to give theoretical explanation to issues of concern. Quantitative data was obtained from respondents who were purposively selected from five private and government owned secondary schools each. Results were analyzed using both descriptive and inferential statistics. This chapter is very revealing, concluding that despite the roles ICT-driven instructional aids has been playing in education, schools in Ogun State, especially those the government owns, are yet to extensively avail and adopt these technologies for teaching and learning processes as a result of certain challenges.

Chapter 6 highlights Potentials of Digital Assistive Technology and Special Education in Kenya by Foad Hamidi, York University, Canada, Patrick M. Owuor, Maseno University, Kenya, Michaela Hynie, York University, Canada, Melanie Baljko, York University, Canada. The authors examine technologies specifically designed for people with disabilities and show how these technologies are paramount in lowering boundaries to education, employment and basic life needs. The also discuss how the growth of a vibrant tech sector in Kenya has had little effect on the prevalence of digital assistive technology in the country. In this most exciting chapter, the authors report on initial explorations undertaken in Kisumu, Kenya to identify existing strengths, relationships, and gaps in access to digital assistive technology. The goal was to explore opportunities for initiatives in participatory design of assistive technology, using an international community/academic partnership. Relevant literature and projects from the areas of Information and Computer Technology for Development (ICT4D), Human-Computer Interaction (HCI) and Critical Disability Studies are reviewed and, these theories are grounded in the authors' experience working with stakeholders in the region. The conclusion discusses promising future directions for participatory and collaborative research in Kenya, and more broadly in the East African context.

Chapter 7 looks at Integrating Digital Literacies Into an Undergraduate Course: Inclusiveness Through the Use of ICTs by Meke I. Kapepo, University of Cape Town, South Africa and Tabisa Mayisela, University of Cape Town, South Africa. In this chapter, the authors share a successful case of how ICT was used for educational inclusiveness by students from previously disadvantaged backgrounds. This chapter further illuminates how students could be supported in acquiring digital and academic literacy skills, within a discipline context. Lecturers often find it difficult to define digital literacy beyond the use of ICT in a classroom, or understand how students' acquisition of digital literacy skills could be supported or further expanded into students' academic literacy skills. This piece of work therefore, provides insight to academics, researchers and students, on how: (1) ICT could be integrated into the curriculum; (2) digital literacy could be embedded into the curriculum; and (3) students could be supported in acquiring graduate attributes and skills that make them fit for the workplace.

Chapter 8 looks at Cloud Computing Adoption Framework for Developing Countries by Anh D. Ta, Advanced Technology Consulting, USA. The chapter explores Cloud computing adoption strategies for small and medium enterprises (SMEs) and microenterprise that can enhance the economic growth of local industries in developing countries. This focus is based on the recognitions that (1) economic growths of local industries reduce poverty through the creation of new jobs for poor people to earn a living wage and support their families, and (2) there is a lack of research on Cloud computing adoption strategy that addresses the unique constraints of small and medium enterprises (SMEs) in developing countries.

Chapter 9 explores ICT for Development in Nigeria, with a focus on ICT4D 2.0. Specifically, the authors investigate Aligning ICT for Development in Nigeria with ICT4D 2.0 Goals. The authors include Olumide. B. Longe, Caleb University, Nigeria, Adefolarin A. Bolaji, Purdue University Graduate School, USA and Richard Boateng, University of Ghana Business School, Ghana. A number of studies have x-rayed the potentials and impetus for development that can be resultant from a full scale adoption of ICT in Africa in general and particularly in Nigeria. Current challenges relating to infrastructures, cybercrime, and government policies and so on that can obviously slow down the benefits accruable from a virile ICT growth has also been highlighted. However, research also need to really address possible areas of developmental benefit of ICT to rural target groups such as farmers, extension officers, health workers and social workers. At the initial Stage, ICT4D1.0 promoted Internet-enabled activities that resulted in globalization, educational, technological, economic and political advantages. The second phase- ICT4D2.0 is geared towards applying ICT much more creatively and pragmatically to development problems beyond the internet-enabled PC tele centers of ICT4D1.0.

Chapter 10 focuses on Levels of Computer Appreciation Skills Acquired for Sustainable Development by Secondary School Students in Nsukka LGA of Enugu State, Nigeria. The authors include Chijioke J. Olelewe, University of Nigeria, Nsukka, Nigeria and Igboamalu P. Nwafor, University of Nigeria Nsukka, Nigeria. This study examines the level of computer skills acquired by secondary school students in Nsukka Local Government of Enugu State, Nigeria. Specifically the study seeks to ascertain the level of word processing skills, graphic application skills and Internet skills acquired by secondary school students in Nsukka Local Government of Enugu State.

Chapter 11 addresses a pertinent issue to small and medium scale business. The chapter contributed by Babafemi Odusote of Covenant University, Nigeria, proposes a conceptual design for Grid Utility SECOnD Portal for Under-Resourced Business Enterprises in Sub-Saharan Africa. The work aspect profiled in paper entails the conceptual design and usability evaluation of a SECOnD portal which serves as the uniform channel through which targeted under-resourced SMMEs can inexpensively access and use e-Commerce technologies and tools without owning them.

While the book is not necessarily a unique reference on ICTs for Development, it greatly contributes in advancing the discourse. This book will not only be relevant for individual reading, but will make a great addition to libraries across the world. The temptation in the research world of ICTs and Development is to borrow “Western” models, collect data, run and analyze to see how it “fits” in the developing nations’ context. This is certainly not enough. Researchers need to keep going on the field to understand ICTs as it pertains to the realities of developing nations, that remain just an afterthought for researchers in the developed world. Is this playing a blame game for these esteemed colleagues in the developed world? Absolutely not. The authors want to challenge researchers based in the developing world to be just as equally aggressive in researching their countries just like colleagues in the developed have done in order to be where they are today. After all these years of technology adoption in developing world, it is no longer anything new that there is a issue of sustainability. However, just stating and making a lot of noise at the many academic conferences focused on ICTs for Development is not enough. Continues quality research is most needed.

It is our earnest hope that this book will make for an easy and most interesting read for all that come across its content. We highly recommend this book for your graduate students working on their masters and doctorate in ICTs and related disciplines. African studies scholars also stand to gain a great deal by having a copy of this book. We wish you a wonderful and interesting read.

Section 1

E-Governance and E-Democracy: Applications and Case Studies

Chapter 1
Assessment of Levels and Interrelationships of ICT Deployment, Web Readiness, and Web Presence Quality of Nigerian E-Government Websites

Kemi Ogunsola
University of Ibadan, Nigeria

Mutawakilu A. Tiamiyu
University of Ibadan, Nigeria

ABSTRACT

Electronic government (e-government), facilitated through government websites are becoming the fastest delivery modes of government services, as they make services available to users, anywhere, anytime. Previous studies focused on the accessibility or quality of these websites in terms of web readiness and/or web presence quality measures. However, no known study has investigated the interrelationships among web readiness, web presence quality and information communication technology deployment for government services (ICT deployment) in Nigeria. The study used a questionnaire (test-retest Spearman r 0.79 for all construct items) to collect data from a sample of 117 public servants in 20 government agencies, on their knowledge of ICT deployment in their agencies; and a checklist for the content analysis of 207 Nigerian government websites, measuring their web readiness and web presence quality. The study recommended that government agencies should re-align ICT deployment with the information, services and features of their e-government websites.

DOI: 10.4018/978-1-5225-2565-3.ch001

INTRODUCTION

Electronic government (e-government) refers to the use of information and communication technologies (ICT) to provide information and government services to citizens, business organizations and other stakeholders. In other words, e-government entails the development and use of different technology-driven platforms and channels for mediating interactions between the government and the governed (Tiamiyu & Ogunsola, 2008).

E-government has brought improvements and changes in the roles and responsibilities of the government and the governed. Improvements are expected in the efficiency of governments' internal processes and in the quality of their interactions with individuals and businesses (Curtin, Sommer and Vis-Sommer, 2003; Siew & Leng, 2003). It has also helped in reshaping the public sector (West, 2004; Guo and Lu, 2005; Kaan, 2007; Yaghoubi, Haghi & Asl, 2011), and increased government transparency (Wong & Welch, 2004; Akunyili, 2010; Adebiyi, Ayo & Adebiyi, 2010).

The Layne and Lee's (2001) model of e-government development recognizes three or four stages of development of e-government, beginning with the information "publishing" or "catalogue" stage, wherein focus is on the provision, through government websites, of information about governments, their objectives and policies and services, and their officials and documents. This is followed by the transaction stage, wherein focus shifts to the actual provision of government transactions with citizens and other stakeholders. The third stage is the integration stage, wherein government information and services become integrated vertically among different hierarchically-related agencies of government and/or horizontally among agencies and stakeholders at the same levels. The final stage is often referred to as the stage of maturity. E-government maturity emerges when e-government services are put online at a high level of performance in order to help citizens to get the best value from their online interactions with the government (Jupp, 2003). Maturity of e-government implies that citizens can obtain information, as well as interact and complete various transactions with government remotely online without necessarily appearing physically at government offices.

In Nigeria, for instance, increased utilization of ICT for the delivery of government services was emphasized in 1999 by the President Obasanjo administration (Choudrie, Umeoji & Forson, 2012). At that time, almost all government services were provided to citizens and other government agencies through the traditional manual methods. This implies that services were accessed through personal visits of customers to government offices, assisted only by postal mails, radio or analogue television broadcasts and limited land-based telephone communications (Tiamiyu, 2005). Then, the fastest delivery modes of government information dissemination and transfer services to customers were primarily the telegraph, cablegram, facsimile, radio, telephone, and analogue television (Okwueze, 2010). Most Nigerians did not really understand and enjoy the inherent benefits of use of modern information technologies by governments, until after the turn of the century, and particularly until the hitherto frustrating telecommunication sector of the country was liberalized in 2003.

Since the early 2000s, there have been efforts made towards electronic connection between the Nigerian government and its people. Some of these efforts in the direction of e-government led the Nigerian government to launch the National Information Technology Policy in March 2001. The implementation began with the establishment of National Information Technology Development Agency (NITDA), under the Ministry of Science and Technology, to champion development of information technology (IT) in Nigeria and promote the implementation of the national IT policy (Fatile, 2012). This has led to the creation of websites by many government agencies, though most of these. Websites are not much interactive and still in the publishing stage (Mundy and Musa, 2010; Olatokun & Adebayo, 2012). Notable progress towards e-government implementation in Nigeria includes: the Nigerian Customs Assycuda Program; the computerization of resident permit and online application for international passport from the Nigerian Immigration Service; the FRSC speed limits diagnostic device for commercial vehicles; and the production of electronic readable driver's license and number plates (Ogunsola, 2015; Ojebola, 2016).

Among the various transaction services that are provided in the maturity stages of e-government are tax declarations, job search in or through assistance by government agencies, application for personal identification documents, such as national passport and driver's license. Other services that can be provided online are car registration, reporting of theft to the police, request and issuance of marriage certificates and birth certificates, enrolment for services provided by public libraries, education and health-related services, registration of new companies, and bidding for government contracts (Atkinson & Leigh, 2003; Liikanen, 2003), among others.

Maturity of e-government services requires the effective and efficient operational deployment of ICT in ways that bring about better interaction, quicker communication, reliable availability and easier accessibility of these services, and integrated seamless access to all government services through a single web portal. In turn, these maturity indicators require interactive and user-friendly government websites for citizen services and reliable back-end infrastructures, and support human resources. This study, therefore, assessed the levels and interrelationships of ICT deployment, web readiness and web presence quality of Nigerian e-government websites.

BACKGROUND

Government agencies in this digital age are now required to focus on providing faster, better, and seamless services, especially through the Internet and other modern ICTs. E-government is described and defined in various ways by different authors, but three things are found to be central to most of these definitions: government, citizens, and the technology used for the interactions between the governments, and the citizens, although, definitions often vary slightly depending on the value, goal and culture of the organizations providing the definitions (Guo & Lu, 2005). Table 1 categorizes some of these definitions.

Table 1. Conceptualizations and definitions of e-government

Categories of Definitions	E-Government Main Target	E-Government Goal	Examples of Authors and Their Definitions of E-Government
First category	Public Administration	Transforming Public Service + Democratic processes	Aicholzer and Schmutzer (2000): e-government is transformation of the business of government, improving service quality delivery, reducing cost, renewing administrative process and transformation of governance itself and re-examining the functioning of democratic practises and processes. Kannabiran, Xavier and Anantharaaj (2005): e-government refers to intensive utilization of information and communication technologies (ICT) for the purpose of forming public intent and rendering services at the political and the administrative level.
Second category	Internal Users + external Users (i.e government administrators and citizens)	Transforming public administration + Services delivery to the citizen	Guo and Lu (2005): e-government is transforming government administration, information provision and service delivery, by the application of new technologies. World Bank (2009): e-government refers to the use by government agencies of information technologies (such as Wide Area Networks, the Internet, and mobile computing) that have the ability to transform relations with citizens, businesses, and other arms of government. Srivastavaan and Teo (2007): e-government is the use of ICT and the Internet to enhance the access to and delivery of all facets of government services and operations for the benefit of citizens, businesses, employees, and other stakeholders.
Third category	Entire society	Development of an information society.	National Academy of Public Administration (2001): e-government is government born of and suitable for an information society– a society increasingly digital and global in its interaction.

The use of e-government websites as a medium of communication and interaction among government, citizens, business organizations and other stakeholders dates back to the 1990s in developed countries. The United States (U.S.) government did a lot of pioneering work on e-government in the 1990s. It started from defining what e-government should be and how it would be achieved in 2001. This translated into the vision published as *E-government: the next American Revolution,* through the effort of the Council for Excellence in Government (McGinnis, 2003). As at 2001, the US web portal FirstGov.gov was at the integrated stage of e-governance (Hiller and Belanger, 2001). The integrated stage of e-government refers to when government services are provided via a single entry point for all government transactions and processes at the federal/national, state/provincial and local government levels (Layne & Lee, 2001; Guo & Lu, 2005).

The U.S. Department of State (http://www.state.gov/) has long incorporated the use of advanced features into government-citizen interactions on its websites, for example, the use of audio and video content of press briefings, as well as providing a video archive that dates back to September 2001. Really Simple Syndication (RSS) feeds are available on the website, including updates to the background notes, country reports, travel advisories, audio and video pod casts (Zammarelli, 2008). The Canadian e-government initiative is unique. This is because of its "one-stop portal", that is, Government On-Line, which provides access to over 130 most common government services online (Accenture, 2004).

In contrast, in most developing countries, such as Nigeria, most government services are still being conducted in the traditional way, involving direct face-to-face interaction between citizens and government officials within the walls of government offices. However, consideration for e-government adoption is gradually becoming priorities for government agencies even in developing countries. South Africa, for example, has adopted additional ways of creating online access to government services, through such channels as Multi-Purpose Community Centers (MPCCs), which is a cooperative effort among all three levels (national, provincial and municipal) of government, as well as non-governmental organizations (Accenture, 2005).

Many other developing countries have started registering their online presence and some have put many of their government services online. But there are various challenges which are peculiar to the developing countries, such as inadequate in-house ICT staff or loss of such staff to the private sector after their recruitment and training, lack of adequate ICT infrastructure, and low Internet access for both the citizens and employees of government agencies. But people in the developing countries are now looking forward to easily accessible government structures and services that respond to their digital age needs and interests (Kannabiran et al., 2005). This is particularly so of the younger generation who are mostly digital natives, have been experiencing the ease of service associated with ATMs, and are increasing owning and using mobile devices to access Internet-based services, social networks, search engines and online stores.

E-government has both supply and demand aspects. The supply side involves what government agencies do to ensure that their services are gradually or quickly transferred to digital modes of delivery, including automated-onsite, phone, online, website and other Internet-based channels and technologies. The demand side involves what citizens, business organizations and other stakeholders and customers of government services do to avail themselves of the services through these channels and technologies.

On the supply side, the provision of government information and services through interactive user-friendly government websites is often emphasized in terms of the capability to save time, promote confidentiality, as well as provide accessibility in more convenient ways (Adebiyi et al., 2010). This is because government websites are becoming online repositories or "libraries" of information intended to disseminate timely and accurate information as well as to promote government-citizen interactions (ActiveStandards & WelchmanPierpoint, 2012).This implies that effective e-government has the potential to bring government closer to the people, make government services more flexible and readily available to citizens, thereby making life more comfortable for them. There is also potential to integrate government processes across agencies, thereby improving inter-agency services and reducing costs (Jupp, 2003; Yaghoubi, Haghi & Asl, 2011).

The clear implications of all these expectations and viewpoints are that, in order for e-government to be successful and sustainable, governments and their agencies must ensure that their e-government websites are adequately geared towards facilitating efficient government-citizen communication, transactions and interactions. They should also provide relevant back-end ICT infrastructure and supportive resources to enhance the quality of e-government services delivered through such websites. In other words, government agencies need to pay attention to various e-government supply-side variables, including those focused on in this study, which are levels of ICT deployment for services by government agencies, the web readiness of their websites, and the web presence quality of the websites.

This study is an update on an aspect of a much larger study (Ogunsola, 2015), which investigated both e-government supply and demand side variables in Nigeria. In this study, ICT deployment by a government agency is defined and measured in terms of the use of ICT equipment and infrastructure, as well as supporting data and human resources and policies by an agency in performing its internal operations and/or providing ICT-based services either on its premises, or through telecommunications and online channels. Thus, the indicators of ICT deployment in an agency include the types of quantities of ICT equipment and infrastructures that are available and their quantities and adequacy, channels and modes of communication with or access to the agency and its information and services that are enabled for the public, internal ICT policies, and availability of information and human capital that supports ICT use in the agency.

Web readiness, in this study, is the preparedness of a government website for the delivery of government information and services. It entails the assessment of a website for provision of e-services, and basic features that can improve government services delivery. The following features were looked out for: functional e-mail addresses of government officials and that of the webmaster; office phone numbers; employment information; links to other government agencies, news materials, government statistics, downloadable forms and government publications. In other words, web readiness refers to the preparedness by a government agency to provide basic information on its website to promote visits to and functional use of the website by its citizens to enquire about, request for, gain access to, or provide feedback on services provided by the agency.

The United Nations Online Network in Public Administration and Finance (UNPAN) has been involved in carrying out e-readiness assessment of the United Nations member countries for many years. The UNPAN assessments provide measured indices for each country for the three defined components of e-readiness – web readiness, telecommunication infrastructure, and human capital. This study, however, focused only on web readiness because data relating to the telecommunication infrastructure index and human capital index are not available for state and local governments or specific government agencies in Nigeria, and could not be collected within the scope and constraints of the study. According to the United Nations (2008), each country's web readiness index was calculated by allocating binary values to the presence or absence of several specific e-government services on the national government's websites.

This study also recognized that it is not only essential that governments deploy services using various ICTs, and websites in particular, but also that the modes of delivery of the services through the websites are of high quality. Thus, the quality of e-government websites was identified in this study as also important to the effective deployment of e-government through websites. Seethamraju (2006) conceives quality of service as the "fitness for use" of a website as perceived by users. In this study, quality of e-government websites and services is referred to as web presence quality. It was measured by ascribing scores to various features of each website, such as multi-lingual features; Web 2.0 tools (blogs, podcast, Twitter, RSS feeds, and others); security/privacy of information; terms of use; interactive capabilities (chat room, feedback forms, and so on); online help facilities; video/audio files; online payment facilities; access to web government document repositories and databases; search capabilities; and start-to-finish online services.

Furthermore, website presence quality also depends on a website's accessibility accommodation qualities. This usually involves assessing a website on five dimensions: navigation and orientation, text equivalence, scripting, styling, and compliance with Hypertext Markup Languages (HTML) standards. An assessment for navigation and orientation includes finding out from a particular website if there are features that will enhance easy movement from one web page to another, or among items on the same page. Assessment for text equivalence finds out if images are properly used on the website and if there are text descriptions for those items that are not textual. The assessment for scripting and styling finds out if the technique used for writing the codes when designing the website allows easy accessibility and interoperability. The HTML guidelines are to improve interoperability and enable the use of varieties of technology for accessing websites. HTML standards are made up of guidelines to assist how text files are tagged, to enhance consistency in the use of font, colours and graphics on websites. It is assumed that, to a large extent, these will improve the web presence quality of e-government websites.

In order to measure web presence quality, various website evaluation tools are often used, mainly to assess the technical qualities of a website. For example, Jati and Dominic (2009) assessed the websites of five Asian countries (Singapore, Korea, Japan, Hong Kong, and Malaysia) using the website performance tool and webpage speed analyzer online service. ActiveStandards and WelchmanPierpoint (2012) carried out a survey of 43 federal government websites of the United States, measuring their performance in three dimensions: accessibility (that is conformity with Section 508 of the US legislation), usability, and search engine optimization (SEO). Bhattacharya, Gulla & Gupta (2007) assessed the portals of 28 state governments and seven union territories in India. They assessed the websites in terms of information quality and system quality.

Most researchers accept that, although the technical merits of websites is a necessary requirement for promoting and delivering effective e-government (which are aspects of the supply side of e-government), it may not be sufficient, as the intended internal and external users of the website (that is, citizens, businesses, government employees in the owner or other agencies and so on) are the ultimate determiners of the quality of a website. However, although Jati & Dominic (2009) recognized this, they did not involve users in assessing the quality of the websites they studied, only recommending that aspect for future studies.

The usual emphasis on the use of online tools to assess the quality of government websites is probably due to the difficulty of conducting compatible user surveys in the different socio-cultural settings of the so many countries often covered by the studies. However, that emphasis needs to be balanced by adequate focus on demand-side variables of e-government as well, including the attitude, perceptions and opinions of actual users of government websites This was the approach adopted in the larger study (Ogunsola, 2015) from which this chapter is derived. The larger study adopted the socio-technical interaction networks theory (Kling, 2000), which states that ICT cannot be analyzed separately from their users. This also agrees with Heeks (2006), who contend that e-government is a hybrid of IT, people and processes. Socio-technical interaction networks theory (Kling, 2000) also guided the choice of research design as well as the data collection processes for the study. Therefore, levels of ICT deployment were derived from responses of public servants, while levels of web readiness and web presence quality of Nigerian e-government websites were derived from the assessment the websites. The study also adopted the Diffusion of Innovation Theory (DOI), which was used as framework to answer the research questions of the study.

MAIN FOCUS OF THE CHAPTER

E-government initiatives in Nigeria, like in most other developing countries, are at the early stage of development. Although Nigeria has made significant efforts on these initiatives, some of them are still at the pilot/test stage. For example, in the Nigerian transport sector, e-verification of vehicle license is still at the pilot stage. But some other e-government services are gradually becoming popular or mandatory. In the Nigerian education sector, registering for and checking of results of many public examinations, such as that of National Examinations Council (NECO) and the Unified Tertiary Matriculation Examination (UTME), are now only allowed online. In the health sector, drug authentication by consumers at the website of the National Agency for Food and Drug Administration and Control (NAFDAC) is gaining popularity. Although some of these e-government services have been adopted as a matter of necessity, there is still the question of how well government websites are prepared to deliver these services, the quality of such websites, and the level of ICT deployed by government agencies to support such online services through the websites.

Some previous studies have attributed the slow pace of use of e-government services to low level of readiness by government agencies (Tiamiyu & Ogunsola, 2008; Olatokun & Adebayo, 2012), perceived inadequate satisfaction of users (Gronier and Lambert, 2010) and low service quality (Colesca & Liliana, 2008; Al-Ghaith et al., 2010). However, there is no known study that has assessed simultaneously the level of web readiness of government agencies, web presence quality of e-government websites, and ICT deployment of government services, and the possible interrelationship among them.

In view of the lack of studies of the status and possible interrelationships among these supply-side variables in Nigeria, the main aim of this study was to help fill the identified knowledge gap. The specific objectives were cast in the form of the following four research questions:

1. What is the level of ICT deployment for the provision of government services in Nigerian government agencies?
2. What is the level of web readiness of the government agencies to provide e-government services through their websites?
3. What is the level of web presence quality of e-government websites in Nigeria?
4. What are the interrelationships among ICT deployment for services by Nigerian government agencies, web readiness of the websites of the agencies, and the web presence quality of the websites?

METHODOLOGY

The study adopted a survey research design. It used the following as main data collection instruments: a checklist for the content analysis of government websites and a questionnaire administered on a purposively selected public servants in a sample of government agencies. The checklist was used to collect information relating to web readiness and web presence quality of websites of government agencies in Nigeria, while the questionnaire was used to collect information on ICT deployment of government services in some of those agencies to obtain matched data for analysis.

The checklist was developed by the researchers based on analysis of the features of government websites in three benchmarking countries– Canada, Republic of Ireland and South Africa. The selection of these three countries was based on a sampling frame which was the Accenture (2005) e-government maturity ranking of twenty-one countries. The systematic sampling led to the selection of the 1st (Canada), 11th (Republic of Ireland) and 21st (South Africa) ranked countries, which served to represent countries with very high, high and moderate e-government maturities that Nigerian e-government programs and websites can learn from. South Africa, which was in the 21st position on the Accenture (2005) list, was the only African country on the list.

Sample sizes were determined separately for the types of population surveyed in this study (government websites and public servants). According to Gay and Airasian (2000), if the population is around 1500, a sample of 20% should be selected. For this study, the total population of available and accessible Nigerian government websites was targeted for assessment, while 40% of the population of the public servants in purposively selected public agencies who were either ICT or senior administrative staff were used. This is because they were in the best position to give information on their agencies' levels of ICT deployment, although the information could also be given by senior officers in the administrative unit of an agency if the agency does not have an ICT department or ICT officers.

Thirty (30) government agencies were purposively targeted and visited for the survey to collect information for assessing levels of ICT deployment. These agencies were referred to as service points because they provide one or more services directly to citizens and businesses organizations. However, only twenty (20) of these agencies actually approved the participation of their employees in the survey (Appendix). There was an estimated 180 respondents who were heads of IT unit, and senior management-level cadre officers, out of which 117 responded; therefore, the response rate was 65%. A total of 72 (61.54%) of the respondents were females; while 45 (38.46%) were males. On years of experience of the public servants, 84 (72.41%) had 1 to 10 years of experience, while the remaining 32 (27.41%) had 11 to 25 years of experience. One respondent did not indicate year(s) of experience. The distribution of the public servants according to their departments showed that 55 (47.83%) were from the administration department, while 60 (52.17%) were from information technology and related departments. Two of the respondents did not indicate their departments.

Section A of the questionnaire that was designed to collect data for measuring ICT deployment for services in the agencies asked questions relating to the background information on the respondents, such as name of the agency, department/unit, position/designation, salary grade level, years of experience in public service and sex. Section B collected data on ICT deployment for services in the respondents' agencies. Items in this section requested for "yes", "no" or "do not know" responses to questions on the availability of e-government services, general e-readiness variables, ICT environment, information capital, human capital and ICT channels for the delivery of e-government services by the agency. Section C collected data on the perceptions of the public servant on the prospects and challenges of e-government services. However, data from this section were not used in the measurement of ICT deployment, as they were intended to assess the public servants' views on the direction for future development of e-government in Nigeria.

A checklist was used for assessing and measuring the web readiness and web presence quality of the government websites. It had five sections. Section A was used to collect identifying information about each website, such as: name of the agency, the URL and date/time of visit/assessment of the site. Section B contained items that checked for and scored various features on the websites, including if there were downloadable forms, government publications, government news, employment information, government statistics, and link to other government agencies. If a feature was available, the "yes" response was marked by the assessors for the website; if the features were absent, a "no" response was marked. Section C collected information on the web presence quality of the government websites. This was done by checking for advanced features on each government website, including if there were multi-language features, Web 2.0 Tools (Twitter, podcast, RSS feeds, etc), interactive features (site maps, video/audio files, search capability, and so on). If a feature was available, the "yes" response was marked by the assessors for the website; if a feature was absent, a "no" response was marked. Section D collected information for determining the accessibility score of each website, on a scale of 1 to 10. Jati & Dominic (2009) used five diagnostic tools.

In this study, only one diagnostic tool, the Functional Accessibility Evaluator (FAE) was used. The FAE assesses almost all the criteria measured by the five tools used by Jati and Dominic (2009), as well as some other criteria they did not measure. The FAE generated scores were based on a maximum score of 100%, but they were converted by the researchers to values based on maximum score of 10 so that the scores would not overweigh the scores from other sub-sections of the checklist. The FAE, as used in this study, assessed a website using five criteria: navigation and orientation, text equivalent, scripting, styling, and compliance with HTML standards. Navigation and orientation is concerned with how well the following are arranged: titles and headings, sub-headings, navigation bars, form control labels, data tables, default languages, access keys and frames. Text equivalent deals with whether a website has informative images, decorative images and image maps.

Other studies evaluated how well a website conformed to W3C standard. This study, however, used subjective evaluation by human experts of each website in Section E of the checklist. Section E required some experts to provide ratings of their perceptions of clarity/non-technicality of language used on the website, background colours, font colours, type and size, background/text color combination, provision for physically challenged persons, appropriate use of images, consistency from page to page, perceived usefulness and perceived ease-of-use. These qualities were rated by the experts on a 10 point scale and averaged for each website.

All the two hundred and ten (210) websites of Nigerian ministries, agencies and departments (MDAs) considered available at the beginning of the study were targeted for the assessment for web readiness and web presence quality. Web readiness and web presence quality were measured for only 207 of the 210 websites that were actually available for assessment. The shortfall of three websites arose from the following: the FAE could not be used on the website of the National Intelligence Agency, owing to security restrictions at the website; the website of the National Institute for Legislative Studies, which was available at the beginning of the study, was later not available for the assessment; and the Nigerian Civil Aviation Authority had two different websites, but only one of which was eventually assessed.

The assessments of the websites with the checklist were performed by the two researchers and their two research assistants, each of who had at least a master's degree in information science or related disciplines. The assessment of the websites of the benchmarking countries was done by the two researchers, and the results provided the basis for the design of the checklist that was used for the assessment of the Nigerian government websites. Subsequently, the research assistants used the checklist for the assessment of the Nigerian government websites. A random sample of their assessments was cross-checked by one of the researchers. Finally, the Functional Accessibility Evaluator (FAE) software version 1.1, developed by the University of Illinois at Urbana Champaign, was used to generate diagnostic and assessment reports on the Nigerian government websites. Data from the reports were combined with data from the use of the checklist to calculate the web presence quality of the websites.

Various analyses were performed on the data collected from the pilot and test-retest of the questionnaire instrument. Prior to the pilot test and test-retest of instruments, face validity of the instruments was done by the two researchers, as well as two other senior researchers in the field of information science. The pilot test of the questionnaire instrument was aimed at collecting information for assessing both face validity and internal validity of the questionnaire items, that each question was clear and unambiguous to respondents, and that the items adequately measured the variables of the study, and that responses to the items that measure a construct in alternative ways were adequately consistent with one another. The test-retest of questionnaires was to enable an assessment of the reliability of the questionnaire for repeated uses of the questionnaire on the same respondents. In the pilot test of the questionnaire for public servants, 10 respondents were involved from three agencies: Nigeria Customs Service (seven respondents), National Directorate of Employment (two respondents) and National Population Commission (one respondent). Amendments were made to both research instruments based on the findings from the pilot and test-retest of the questionnaire and the face validity of the checklist.

In order to compute the web readiness score for each website, the availability or absence of each web readiness checklist item was coded as '1' and '0' for availability and absence, respectively. For each website, four assessors were involved and the average score for the availability of each feature on the website was calculated. This was in order to compute a summary score for each assessed feature and overall for each website. After the computations, the minimum achievable score for web readiness was 0, while the maximum achievable score was 66. This range was subdivided into five labelled sub-ranges as follows: 0.00 to 13.99("Very low web readiness); 14.00 to 26.99 ("low readiness"); 27 to 39.99 ("moderate web readiness"); 40 to 52.99 ("high web readiness"; and 53.00 to 66 "very high readiness").

Web presence quality measurement was based on assessment of the advanced features on government websites, website ratings by the four experts, and data provided by the Functional Accessibility Evaluator (FAE), with the data contributed by each of these sources converted to 1 to 10 rating scales in order to be able to add them to obtain an aggregate score for web presence quality for each website. This use of ratings scales for website quality was used in other studies, such as in ActiveStandards and WelchmanPierpoint (2012), which measured websites performance on a 1-10 rating scale, with 1 as the lowest score and 10 as the highest. Bhattacharya, Gulla & Gupta (2007) used a 5-point scoring system to grade the portals in terms of information quality and system quality.

RESULTS AND DISCUSSIONS

Research Question 1: What is the level of ICT deployment for the provision of government services in Nigerian government agencies?

For an agency to deploy e-government services, such agency must have good organizational e-readiness for different ICT-enabled service delivery channels (multi-channels), appropriate policy in relation to e-service provision, as well as actually provide a number of such services via the Internet. Yaghoubi, Haghi & Asl (2011) stress the importance of using various channels for e-government development, indicating that governments can only show sensitivity to local realities by assessing and evaluating different technology alternatives. The United Nations (2008) observes that, in many countries, the cornerstone of the e-government strategy is the adoption of multichannel service delivery.

In this study, as explained above, a questionnaire survey of public servants working in the agencies was used to collect data to assess and measure the level of ICT deployment for services in the agencies. For the survey, a total of 117 public servants were sampled from 20 purposively selected key public agencies. The questionnaire contained 37 questions, a few of which had sub-questions, which were categorized in the questionnaire into sub-groups relating to availability of e-services, network connectivity, information capital, human capital and ICT policies. A "yes" response was coded as 1, while "no" and "don't know" were coded as 0. Thus, for the question and sub-questions, an agency could obtain a total score of between 0 and 46 inclusive, and this range was divided into the following five equally-spaced sub-ranges for levels of ICT deployment: 0 to 9.99 ("very low), 10 to 18.99 ("low"), 19 to 27.99 ("moderate"), 28 to 36.99 ("high"), and 37 to 46.99 ("very high").

The results showed that 42 (35.9%) of the respondents answered the different questions in the survey questionnaire which indicated that their agencies had very

Figure 1. Level of ICT deployment of e-government services (N = 20 agencies)

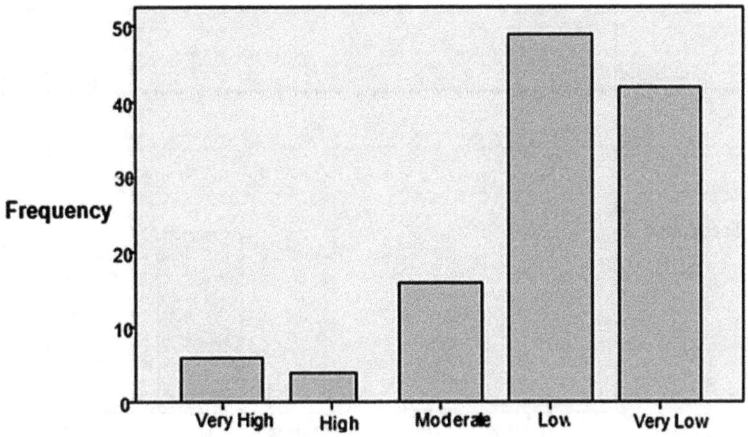

low level of ICT deployment; 49 (41.9%) public servants gave responses which indicated low level of ICT deployment; 16 (13.7%) respondents gave responses that shows moderate level of ICT deployment; 4 (3.4%) respondents gave responses that showed high level of ICT deployment; while 6 (5.1%) respondents provided responses that indicated that their agencies had very high level of ICT deployment (Figure 1).

Research Question 2: What is the level of web readiness of the government agencies to provide e-government services through their websites?

The data for answering this question were obtained through the checklist, which was used to assess 207 government websites, as explained above. The websites that had very low web readiness were 126 (60%); while 66 (31.4%) of them had low levels of web readiness. Further, 14 (6.7%) had moderate web readiness, only one 1 (0.5%) had high level of readiness; and only two (1%) had very high level of web readiness (Figure 2).

The websites of some state governments in Nigeria (4 out of 36) were not available during part of the period of the assessment. These states were those where new governors had just been sworn in and where the outgoing and the incoming governors were from different political parties. As is common in Nigeria and many developing countries where institutions are often personalized by political office holders, the website are often pulled down at such times in order for the incoming governors to approve a new website or new website content to be developed in such a manner that it completely removes information about the former governors. In situations like these, the citizens and businesses suffer as they lose connection with the former website along with the services and projects of the previous administration.

Figure 2. Level of web-readiness of e-government websites (N = 207 websites)

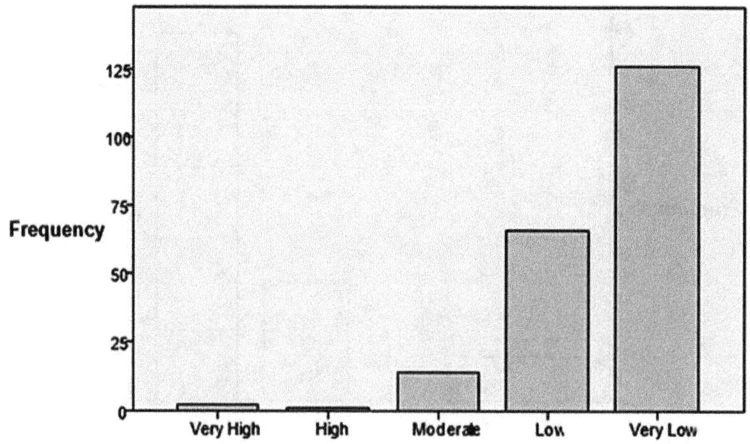

According to Andersen & Henriksen's (2000) Public Sector Process Rebuilding model, as e-government progresses in government institutions, it should grow from the "cultivation" stage (elementary stage) to an advanced stage of "maturity", where emphasis would be on accountability and transparency of government processes, and then on to the stage of "revolution", where the ownership of data and services would be institutionalized and transferred to all stakeholders, including not only political office-holders, but also citizens, business organizations, academic institutions, non-governmental organizations and the civil society.

The low level of web readiness recorded by many of the government agencies is supported by the finding of United Nations (2005; 2008; 2010). The United Nations Web Measure Index showed that the web readiness of Nigerian government websites was low compared to the three purposively selected benchmark countries in the study: Canada, Ireland and South Africa. For instance, in 2005 when 191 member states of the United Nations were accessed, Nigeria's web measure index was 0.2231, that of South Africa was 0.5692, that of Ireland was 0.7115, and that of Canada was 0.8923. Similarly, in 2008, the web readiness for the countries was as follows: Nigeria, 0.2241; South Africa, 0.5518; Ireland, 0.6756; and Canada, 0.7659. The Nigerian government websites achieved a dramatic increase in the web measure index, from a very lowly 0.083 in 2003 to 0.2231 in 2005, but only followed by a minimal additional increase to 0.2241 in 2008. Government websites with very high readiness can be regarded as innovators in the area of web readiness, while the ones with the lowest e-readiness are regarded as the laggards. Therefore, using the parameters of Rogers (1983) and Sahin's (2006) Diffusion of Innovation Theory, it can be said that 60% of the websites were laggards in relation to web readiness;

31.4% were late majority; 6.7% were early majority; 0.5% was early adopters; and only 1% were innovators.

Research Question 3: What is the level of web presence quality of e-government websites in Nigeria?

Out of a total of 207 websites assessed, only one (0.5%) website had "low web presence quality," three (1.4%) websites had "very low web presence quality"; while 13 (6.2%) websites had moderate web presence quality. A total of 100 (47.6%) websites had "high web presence quality"; while 90 (43.3%) had "very high web presence quality" (see Figure 3).

The result indicated that many government websites in Nigeria were becoming websites of high quality. However, at the time of this study, the Nigerian national government had not provided a seamless portal. E-government will thrive if the concept of "many agencies, one government" is integrated from the customer service delivery channels, regardless of the number of agencies involved in providing the services (Siew & Leng, 2003). A smart e-government initiative will solve the problem of scattered and disjointed government agencies websites, since many people are not interested in which government agency or tier of government is responsible for a service, neither are they interested in bureaucratic acronyms (Atkinson & Leigh, 2003). In most cases, what the citizens want is satisfactory government services, at any time and place. This agrees with a study by Kanda, kulthida & Wasu (2014), which found that considering citizens' need for e-government services, quality is their priority. Alshehri, Drew, Alhussain & Alghamdi (2012) affirm that quality

Figure 3. Level of web presence quality of government websites (N = 207 websites)

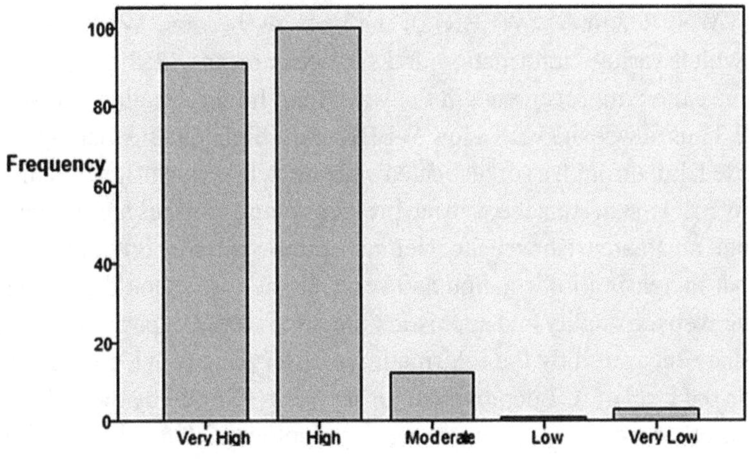

e-government websites, which is one of the important delivery points for e-services, is essential in the successful adoption of e-government services by the citizens.

In this study, websites with a high web presence quality are those with more web features, such as search capability, site map, video and audio files, and social media functions. Such websites would be rated high by both FAE and human experts who assessed the websites. This is because the availability of each of the identified features on each website fetched specified scores for it. Website qualities that were rated and scored by human experts include: clarity/ non-technicality of language used, background colour, font colours, sizes and types, background/text colour combination, provision for physically challenged persons, appropriate use of images, consistency from page to page, perceived usefulness, and perceived ease-of-use. According to the perspective of the Diffusion of Innovation theory, the websites with very high web presence quality are like innovators, the ones with high web presence quality are like the early adopters, websites with moderate web presence quality were like the early majority, those with low web presence quality are like late majority, while those with very low web presence quality are like the laggards. In this study, a total of 91 (43.3%) websites with very high web presence quality were regarded as website at the innovation stage, while only 1(0.5%) website was considered as being in the laggard state.

Research Question 4: What are the interrelationships among ICT deployment for services by government agencies, web readiness of the websites of the agencies, and the web presence quality of the websites?

This research question is the ultimate thesis of this study, whether there are any meaningful interrelationships or interdependencies among the three supply-side e-government variables focused upon in the study. Logically, and as shown in Figure 4, one would expect Web Presence Quality (WEBQ) of a website to be directly related to Web Readiness (WEBR) of the website, because WEBR measures the level to which various information and services are provided on websites, while WEBQ measures improvements in the ways that the information and services are provided. Thus, a website with a low WEBR and a high WEBQ means few services are provided, but through a sophisticated website; while a website with high WEBR and low WEBQ means that the website provides many services and service features, but through an amateurish website. Neither of these is really optimal, as one would expect that increasing information and services on a site should be matched with increasing website quality and sophistication. One would expect the web readiness of an agency measured by the information and services provided at its website to depend on the level of ICT deployment for services (ICTDS) by the website's owner agency. In other words, the more an agency deploys ICT for the delivery of both

Figure 4. Model of the expected or actual interdependence among ICTDS, WEBRD, and WEBPQ

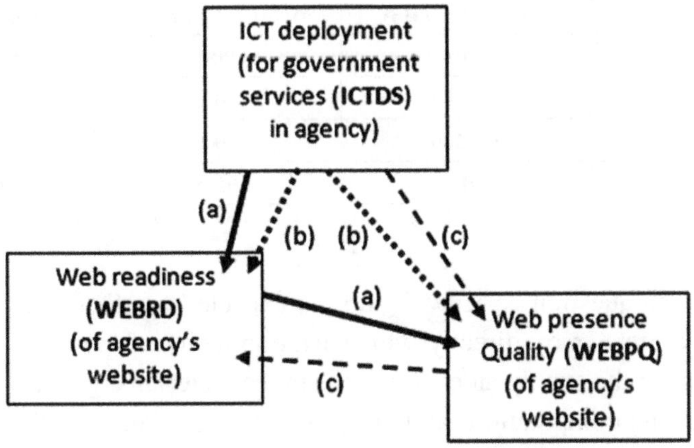

its internal operations and external services, the more one would expect that it will consider the website as one of the key channels for providing some of the services.

In summary, the above line of reasoning, or model, suggests that the more an agency is web-ready (that is provides more and more information and services through its website), the more likely it would have considered and implemented advanced facilities and features on its website, although this may be at a later stage of website development. Thus, as shown by the black arrowed line labelled (a) in Figure 4, this model postulates that high web readiness is the expected logical consequence of higher ICT deployment for services (ICTDS) initiatives, and that high web presence quality (WEBPQ) is the expected logical follow-up to high web readiness (WEBRD). However, other interdependencies among these three variables are possible in reality. Two of these other possibilities, as shown by the hashed and dotted lines labelled (b) and (c) in the Figure 4, are:

(b) ICT deployment leading directly to both high web readiness and high web presence quality through a sophisticated website. This mode requires simultaneous ICT deployment along with synchronized website development to a high standard in both web readiness and web presence quality terms.

(c) some initial web-oriented ICT deployment birthing a sophisticated website with high web presence quality, but lacking adequate web readiness in terms of the provision of customer-centric information and services, possibly followed later by further ICT deployment investments to remove these initial deficiencies. However, the initial ICT deployment might not be sustained or sustainable, resulting in a sophisticated but dysfunctional website, that is not maintained.

Table 2. Descriptive statistics for ICTDS, WEBRD, WEBPQ

	ICT Deployment for Services (ICTDS)	Web Readiness of Websites (WEBRD)	Web Presence Quality of Websites (WEBPQ)
N	20	207	207
Mean	30.7600	14.0934	85.7608
Median	30.2500	13.0000	87.7656
Std. deviation	9.0142	8.95152	12.2388

The research question is: How far do the expected or other interdependencies reflect reality? Stated more directly: which of the above interdependencies or possibly others are true of the supply-side conditions in the Nigerian e-government context?

The following computations and analyses were performed on the data in order to answer this question. Firstly, the responses of the public servants from the same agency who participated in the questionnaire survey to assess ICT deployment for services (ICTDS) were averaged to obtain a summary ICTDS value for the agency. Secondly, normality tests were performed on the data on each of the three variables – levels of ICT deployment for services (ICTDS), levels of web readiness (WEBRD), levels of web presence quality (WEBPQ), to ascertain that they were approximately normally distributed. These tests confirmed this for all three variables.

Thereafter, the following analyses were performed. Descriptive statistics were produced for the three variables (Table 2). These statistics showed that, comparatively, variability of WEBPQ (web presence quality) was low among the websites (standard deviation only about one-seventh of the mean), followed by the variability among the agencies in their ICT deployment for services (ICTDS) (standard deviation about one-third of the mean), while variability was high among the websites on web readiness (WEBRD) (standard deviation more than half of the mean).

Next, Pearson correlation analyses were performed among the three variables for the twenty agencies for which data on levels of ICT deployment for services (ICTDS) were available. The results are shown in Table 3. Finally, Pearson correlation analysis was again done between the measured levels of web readiness (WEBRD) and levels of web presence quality (WEBPQ) of the 207 agencies for which data on these two variables were available, in order to confirm the earlier Pearson correlation results in Table 3 between the two variables (Table 4).

The information in Table 3 shows that, whereas the web readiness (WEBRD) and web presence quality (WEBPQ) of the government websites were strongly positively correlated ($r = .581$, $p = .007$) for the 20 agencies for which corresponding data were available, there was no correlation between ICT deployment for services (ICTDS) and each of the other two variables ($r = -.019$, $p > .05$; and $r = .027$, $p >$

Table 3. Statistics for Pearson correlation of ICTDS, WEBRD and WEBPQ (20 agencies)

Variable	ICTDS	WEBRD
ICTDS (ICT deployment for services)		
WEBRD (Web readiness of website)	r = -.019 p =.936 N = 20	
WEBPQ (Web presence quality of websites)	r =.027 p =.910 N = 20	r =.581 p =.007 N = 20

N - No of cases (agencies);
r – Pearson correlation coefficient;
p – 2 tailed significance.

Table 4. Statistics for Pearson correlation between WEBRD and WEBPQ (207 websites)

Variable	WEBPQ (Web Presence Quality of Website)
WEBRD (Web readiness of website)	r =.329 p =.000 N = 207

N - No of cases (websites);
r – Pearson correlation coefficient;
p – 2 tailed significance.

0.05 respectively). The very strong correlation between web readiness (WEBRD) and web presence quality (WEBPQ) of the government websites was confirmed again when data on the two variables for the 207 websites were correlated (Table 4), where r =.329, p =.000. It is, however, also instructive to stress that the 20 agencies for which data were available on all three variables are the key agencies of federal government of Nigeria which were purposively selected for the survey of ICT deployment, web readiness and web presence quality.

What these results point to is that the provision of information and services, captured by the web readiness (WEBRD), and the provision of advanced features, captured by the web presence quality (WEBPQ) of the websites, do not correlate with what the website owner agencies are doing internally to deploy ICT for the delivery of their services, as captured by ICT deployment for services (ICTDS). The results also suggest that there might be disconnection between ICT deployment for services initiatives in the agencies and the nature of the information, services and advanced features provided by their websites, and that this disconnect might affect the ability of the agencies to effectively drive, support or sustain the information, services and advanced features promised or provided by their websites. This is against the ideal e-government service provision situation, whereby ICT deployment for services within a government agency (that is the back-end infrastructure and policies) are harmonized and synchronized strategically with the information, services and features on the front-end websites and other channels.

Now, how do these results fit in with the model in Figure 4? Recall that the findings in respect of research questions 1, 2 and 3 reported above are that: most of the government agencies scored low on ICT deployment and web readiness, but high on web presence quality. These findings, taken along with the patterns of correlations between pairs of the variables seem to point to the following conclusion: most of the Nigerian government agencies surveyed in this study had advanced features on their websites, but the quality of actual information and services provided to their customers on the websites are poor, in addition to poor levels of ICT deployment for services by the agencies themselves. The inevitable conclusion is that Nigerian agencies are often trying to run with fanciful features on their websites (as most likely dictated by their usually outsourced website designers) before learning to walk on the websites with basic customer-oriented information and services that depend crucially on ICT deployment for services by the agencies. This disconnection between ICT deployment in the agencies and the web readiness and web presence quality of their websites suggests that the agencies might not yet have recognized the required levels of backend ICT infrastructure and associated supportive resources that are needed to provide effective and customer–satisfying online services through their websites.

SOLUTIONS AND RECOMMENDATIONS

The recommendations to government and government agencies in Nigeria arising from the insight gained through this study are:

1. Government agencies should work to ensure that their websites are not just publishing any information, but that the nature of information published is adequately customer-centric in addition to the current emphasis on providing agency-politics and politician-centric information.
2. The websites should also be upgraded in order for them to be adequately transactional and interactive in order to improve their web readiness ratings.
3. There is a need for more integrated government web portals (a single entry point for all government transactions and processes) at the federal and state government levels, thereby moving government agencies from the era of unlinked websites or disjointed e-government portals to that of government

online gateways for accessing varied government services provided by various agencies.

4. There is a need to implement website improvement policies and enhancements that meet the special needs of people living with disabilities. This is because most of the assessed Nigerian government websites did not take into account the special needs of people living with disabilities to support their use of the e-government services.
5. Nigerian government agencies need to gear up to rectify the observed disconnection between levels of their ICT deployment for their services and the web readiness and web presence quality of their websites. This can be achieved by investing adequately in ICT deployment initiatives and properly synchronizing the information, services and features provided by their websites with their adequately organized back-end ICT infrastructure and other resources.

FUTURE RESEARCH DIRECTIONS

An examination of the important government websites that came up after data collection for this study could be done. Examples are: Services.gov.ng (http://services.gov.ng/), which was developed 2013-2014; and Federal Republic of Nigeria government website (http://www.nigeria.gov.ng/index.php), which was under reconstruction at the time of data collection for this study (2012) but was later made available in 2014. Other important Nigerian government websites could also be included in future studies. Furthermore, studies should be undertaken on the effect of the ICT deployment, web readiness, and the web presence quality of Nigerian government websites on the customer use of and satisfaction with e-government services in Nigeria.

CONCLUSION

The findings of this study showed that the Nigerian government websites had sophisticated features that conform to current global good practices in website design, but the basic information and services provided on the sites were inadequate. The study also revealed that the level of supportive back-end ICT deployment for services was low. Moreover, at the time of this study, the Nigerian federal and state governments lacked seamless national e-government portals. Four out of the thirty-six states had no functioning or accessible official websites. Only nine out of the 744 local governments had websites, and only 169 out of the 488 MDAs had functional websites. Clearly, Nigerian governments and their agencies need to deploy more ICT and improve on the readiness of their websites for effective and revolutionary e-government.

ACKNOWLEDGMENT

The authors gratefully acknowledge the sponsorship and support of the Social Science Research Council (with funds from Carnegie Corporation, USA); John D. and Catherine T. MacArthur Foundation, USA; Centre for Technology in Government, University at Albany, City University of New York, USA; and Faculty of Information and Media Studies (with funds from Canadian Bureau for International Education on behalf of Department of Foreign Affairs and International Trade) which made this project and its outcomes possible.

REFERENCES

Accenture. (2004). *E-government leadership: high performance, maximum value.* Retrieved September 2, 2006, from http://www.accenture.com/Global/Research_and_Insights/By_Industry/Government/HighValue.htm

Accenture. (2005). E-*government leadership: leadership in customer service: new expectations, new experiences.* Retrieved September 2, 2006, fromhttp://www.accenture.com/Countries/Canada/Services/By_Subject/Customer_Relationship_Management/R_and_I/Le adershipNewExperiences.htm

Active Standards & Welchman Pierpoint. (2012). *US e-government website quality report.* Retrieved November 14, 2013, from www.welchmanpierpoint.com

Adebiyi, A. A., Ayo, C. K., & Adebiyi, M. O. (2010). Development of electronic government procurement (e-GP) system for Nigeria public sector. *International Journal of Electrical & Computer Sciences, 10*(6), 69–76.

Akunyili, D. (2010). *ICT and e-government in Nigeria: opportunities and challenges.* World Congress on Information Technology, Amsterdam, The Netherlands. Retrieved April 13, 2011, from www.goafrit.wordpress.com/2010/06/12/ict-and-e-government-in-Nigeria

Al-Ghaith, W., Sanzogni, L., & Sandhu, K. (2010). Factors influencing the adoption and usage of online services in Saudi Arabia. *The Electronic Journal on Information Systems in Developing Countries, 40*(1), 1-32. Retrieved September 3, 2013, from http://www.ejisdc.org

Alomari, M. K., Woods, P., & Sandhu, K. (2009). The deployment of e-government in the Hashemite Kingdom of Jordan: factors in e-government adoption.*IADIS International Conference WWW/INTERNET*.

Alshehri, M., Drew, S., Alhussain, T., & Alghamdi, R. (2012). The effect of website quality on adoption of e-government services: An empirical study applying UTAUT Model Using SEM. *23rd Australasian Conference on Information Systems*.

Andersen, K. V., & Henriksen, H. Z. (2000). The first leg of e-governance research: Domains and application areas 19982003. *International Journal of Electronic Research, 1*(4), 26–44. doi:10.4018/jegr.2005100102

Atkinson, R. D., & Leigh, L. (2003). Customer-oriented e-government: can we ever get there? In The World of E-government. New York: The Haworth Press, Inc. doi:10.1300/J199v02n03_10

Bhattacharya, D., Gulla, U., & Gupta, M. P. (2007). *Assessing effectiveness of state government portals in India.* Computer Society of India. Retrieved December 17, 2013, from www.csi-siegov.org

Choudrie, J., Umeoji, E., & Forson, C. (2012). *Diffusion of e-government in Nigeria: a qualitative study of culture and gender.* University of Hertfordshire Business School Working Paper. Retrieved August 12, 2012, from https://uhra.herts.ac.uk/dspace/handle/2299/5549

Colesca, S. E., & Liliana, D. (2008). E-government adoption in Romania. *World Academy of Science, Engineering and Technology, 42*, 140–145.

Creswell, J. W. (1994). *Qualitative and quantitative methods.* Newbury Park, CA: Sage.

Curtin, G. G., Sommer, M. H., & Vis-Sommer, V. (Eds.). (2003). The world of e-government. New York: The Haworth Press, Inc.

Fatile, J. O. (2012). Electronic government: Myth or opportunity for Nigerian public administration? *International Journal of Academic Research in Business and Social Sciences, 2*(9), 122–140.

Gay, L. R., & Airasian, P. (2000). *Educational research: competence for analysis* (6th ed.). New Jersey: Merrill.

Gronier, G., & Lambert, M. A. (2010). A model to measure the perceived quality of service in e-government.*Proceedings of 10th European Conference on E-government*. University of Limerick.

Guo, X., & Lu, J. (2005). Effectiveness of e-government services in Australia. In Electronic Government Strategies and Implementation. Hershey, PA: Idea Group Publishing. doi:10.4018/978-1-59140-348-7.ch009

Heeks, R. (2006). *Implementing and managing e-government*. London: SAGE publications.

Hiller, J. S., & Belanger, F. (2001). Privacy strategies for electronic government, in e-government. Oxford, UK: Rowman and Littlefield Publishers.

Jati, H., & Dominic, D. D. (2009). Quality evaluation of e-government website using Web Diagnostic Tools: Asian case. *2009 International Conference on Information Management and Engineering*. Retrieved November 15, 2013, from http://ieeexplore.ieee.org/

Jupp, V. (2003). Realizing the vision of e-government. In The World of E-government. New York: The Haworth Press, Inc.

Kaan, K. A. (2007). *Online usability among Arizona local governments* (Ph.D. dissertation). Arizona State University. Retrieved January 16, 2009, from http://proquest.umi.com/

Kanda, S., Kulthida, T., & Wasu, C. (2015). Factors affecting the development of e-government using a citizen-centric approach. *Journal of Science & Technology Policy Management*, *6*(3), 206–222. doi:10.1108/JSTPM-05-2014-0027

Kannabiran, G., Xavier, M. J., & Anantharaaj, A. (2005). Enabling e-governance through citizen relationship management-concept, model and applications. *Journal of Service Research*, *4*(2), 223–240.

Kling, R. (2000). Learning about information technologies and social change: The contribution of social informatics. *The Information Society*, *16*(3), 217–232. doi:10.1080/01972240050133661

Kumar, V., Mukerji, B., Butt, I., & Persaud, A. (2007). Factors for successful e-government adoption: A conceptual framework. *The Electronic. Journal of E-Government*, *5*(1), 63–76.

Layne, K., & Lee, J. (2001). Developing fully functional e-government: A four stage model. *Government Information Quarterly*, *18*(1), 122–136. doi:10.1016/S0740-624X(01)00066-1

Liikanen, E. (2003). E-government: an EU Perspective. In The World of E-government. New York: The Haworth Press, Inc.

McGinnis, P. (2003). Creating a blueprint for e-government. In The World of E-government. New York: The Haworth Press, Inc. doi:10.1300/J199v02n03_04

Mundy, D., & Musa, B. (2010). Towards a framework for e-government development in Nigeria. *Electronic. Journal of E-Government*, *8*(2), 148–161.

National Academy of Public Administration. (2001). *The e-government manifesto*. Retrieved April 2, 2010, from www.napawash.org/cim.nfs?OpenDatabase

Ngulube, P. (2007). The nature and accessibility of e-government in sub-Saharan Africa. *International Review of Information Ethics*, *7*(9), 1–10.

Ogunsola, K. (2015). *Predictors of use of electronic government services in Nigeria* (Unpublished doctoral thesis). University of Ibadan, Ibadan, Nigeria.

Ojebola, T. (2016). *Users' assessment of the quality of information disseminated through facebook by selected law enforcement agencies in Nigeria (Master's degree project)*. Africa Regional Center for Information Science University of Ibadan.

Okwueze, F. O. (2010). E-Governance as a tool for public sector development in Nigeria. *International Journal of Research in Arts and Social Sciences,* 2. Retrieved January 29, 2014, from http://www.academicexcellencesociety.com/journals_ijrass.html

Olatokun, W. M., & Adebayo, B. M. (2012). Assessing e-government implementation in Ekiti State, Nigeria. *Journal of Emerging Trends in Computing and Information Sciences*, *3*(4), 497–505.

Rogers, E. M. (1983). *Diffusion of innovations* (3rd ed.). New York: The Free Press, Macmillan Publishing, Inc.

Sahin, I. (2006). Detailed review of Rogers' diffusion of innovations theory and educational technology-related studies based on Rogers' theory. *The Turkish Online Journal of Educational Technology*, *5*(2), 14–23.

Seethamraju, R. T. (2004). Measurement of user-perceived web quality.*Proceedings of the 13th European Conference on Information Systems.*

Siew, L. S., & Leng, L. Y. (2003). E-government in action: Singapore case study. In The World of E-government. New York: The Haworth Press, Inc.

Stowers, G. (1999). Become cyberactive: State and local governments on the World Wide Web. *Government Information Quarterly*, *16*(2), 111–127. doi:10.1016/S0740-624X(99)80003-3

Tiamiyu, M. A., & Ogunsola, K. (2008). Preparing for e-government: Some findings and lessons from government agencies in Oyo State, Nigeria. *South African Journal of Library and Information Science*, *74*(1), 58–72.

Tiamiyu, M. A. (2005). *Using Information Technology for Total Quality Management of Government Business in Nigeria*. Akure, Nigeria: ITCC Training Course.

United Nations. (2004). *Towards access for opportunity*. Retrieved February 10, 2007, from unpan3.un.org/egovkb/en-us/Reports/UN-E-Government-Survey-2004

United Nations. (2005). *From e-government to e-inclusion*. Retrieved February 10, 2007 from unpan3.un.org/egovkb/en-us/Reports/UN-E-Government-Survey-2005

United Nations. (2008). *From e-government to connected governance*. Retrieved February 10, 2009, from http://unpan1.un.org/intradoc/groups/public/documents/UN/UNPAN028607.pdf

United Nations. (2010). *World e-government rankings*. Retrieved October 21, 2011, from http://kz.mofcom.gov.cn/accessory/201009/1284225105383.pdf

UNPAN. (2004). *UN Global E-government Readiness Report 2004*. Retrieved February 10, 2009, from http://unpan1.un.org/intradoc/groups/public/documents/un/unpan019207.pdf

West, D. M. (2004). E-government and the transformation of service delivery and citizen attitudes. *Public Administration Review*, *64*, 15–27.

Srivastava, S. C., & Teo, T. S. H. (2007). E-government payoffs: Evidence from cross-country data. *Journal of Global Information Management*, *15*(4), 20–21.

Wong, W., & Welch, E. (2004). Does e-government promote accountability? A comparative analysis of website openness and government accountability. governance. *An International Journal of Policy, Administration, and Institutions*, *17*(2), 275–297. doi:10.1111/j.1468-0491.2004.00246.x

World Bank. (2009). *Definition of e-government*. Retrieved March 9, 2009 from http://go.worldbank.org/M1JHE0Z280

Yaghoubi, N. M., Haghi, A., & Asl, S. K. (2011). E-government and citizen satisfaction in Iran: Empirical study on ICT offices. *World Applied Sciences Journal*, *12*(7), 1084–1092.

Zammarelli, C. M. (2008). *Advanced content in state e-government: criteria for evaluation* (Master's degree thesis). Graduate School, University of Maryland, College Park. Retrieved January 16, 2009, from http://proquest.umi.com/

ADDITIONAL READING

Becker, S. A. (2008). Accessibility of federal electronic government. Chen H., Brandt L., Gregg V., Traunmuller R., Dawes S., Hovy E., Macintosh and Larson C. Eds. Digital Government: E- Government Research, Case Studies and Implementation. New York: Springer doi:10.1007/978-0-387-71611-4_8

West, D. M. (2005). *Global e-government. Digital Government: Technology and Public Sector Performance*. Princeton University Press.

KEY TERMS AND DEFINITIONS

E-Government: The use of ICT for the provision of adequate information and government services to citizens and business organizations irrespective of time and/or location, and the creation of platforms for interactions between the government and the governed.

ICT Deployment: The availability of e-services provided online and mode of communication with the services, IT policy environment, information capital, human capital, and various ICT channels for the delivery of e-government services.

Web Presence Quality: The process of enumeration and evaluation of government websites and thereby ascribing scores for the features that will increase websites' friendliness towards users.

Web Readiness: The preparedness of a government website for the deployment and use of e-government services by enumerating basic features that can improve government services delivery, such as e-mail address of government officials, e-mail address of the webmaster, and links to other government agencies.

APPENDIX: LIST OF NIGERIAN GOVERNMENT MINISTRIES, DEPARTMENTS AND AGENCIES (MDAS) VISITED FOR THE SURVEY OF PUBLIC SERVANTS TO ASSESS THEIR LEVEL OF ICT DEPLOYMENT FOR THEIR SERVICES

1. Abuja Geographic Information System (AGIS)
2. Bureau of Public Procurement (BPP)
3. Corporate Affairs Commission (CAC)
4. Federal Inland Revenue Service (FIRS)
5. Federal Ministry of Health
6. Federal Ministry of Information
7. Federal Mortgage Bank of Nigeria
8. Federal Road Safety Corp (FRSC)
9. Joint Admission and Matriculation Board (JAMB)
10. Lagos State Government
11. National Agency for Food, Drug Administration and Control (NAFDAC)
12. National Examination Council (NECO)
13. National Identity Management Commission (NIMC)
14. National Open University Of Nigeria (NOUN)
15. Nigeria Customs Service (NCS)
16. Nigeria Export Promotion Council (NEPC)
17. Nigeria Immigration Service
18. Nigerian Civil Aviation Authority (NCAA)
19. Security and Exchange Commission (SEC)
20. Standards Organisation of Nigeria (SON)

Chapter 2
Interrogating the Application of E-Governance for Service Delivery:
A Study of Ojo and Alimosho Local Government Areas, Lagos State

Francis Friday Nchuchuwe
Lagos State University, Nigeria

ABSTRACT

This study examined the application of e-governance for service delivery in Ojo and Alimosho Local government areas of Lagos State. Questionnaire and Interviews were used as major instruments for data collection alongside other secondary sources. The simple Random Sampling technique was used to derive a sample of 100 respondents from a population size of 748 workers from Ojo and Alimosho Local Government Areas. Simple percentages were used for the demographic data analysis, as well as to test the hypotheses. The findings revealed that the application of E-governance is still a far cry in the local governments. Generally, the concept is novel among the workers as it is generally construed to mean electronic payment of salaries and lodgments into the Banks by the citizens. Beyond this, it was revealed that such facilities like web sites, constant electricity supply, enabling environments to boost tele-density, internet diffusion, adult computer literacy, cyber cafes, computer, etc. were not adequately available as expressed by the respondents.

DOI: 10.4018/978-1-5225-2565-3.ch002

INTRODUCTION

The application of ICT at every facet of human's existence, the world over has reached extraordinary heights, stimulating fundamental changes and development in the way people and organizations relate, share information and conduct transactions (Nchuchuwe & Ojo, 2015). This probably informed why Edewor, Imhonopi & Urim (2014) pointed out that Information and Communication Technology (ICT) has become, within a very short time, one of the basic building blocks of modern society. According to them, ICTs have successfully changed the social, economic and political spaces globally. In Nigeria various reforms have been carried out and policies implemented with the aim of positioning the local government system in Nigeria for quality service delivery. The enactment of the National Information Technology policy and National Information Technology Development Act of 2001 and 2007 respectively were among the moves by the federal government to apply ICT for service delivery to the citizens. However, despite these efforts the application of ICT is still a far cry at all levels of government. According to Arowolo (2012), Performance of Nigerian public service has been a major concern to policy makers and researchers as well. This is because despite all measures put in place to arrest the performance failure, the service, it seems, has defied all approaches towards tackling the problem of inefficiency and capacity collapse. The submission sums up the perennial problems bedeviling public service delivery in Nigeria. The Local government is not an exception. Indeed, not only is service delivery poor, the political elites have completely dominated the policy and decision making process thereby denying the populace the right to participate in the decisions that affect them especially for service delivery. Coupled with this is poor revenue generation and colossal corruption in the local governments. As Adeyemi (2012) puts it, "corruption in the local government has become so canonically accommodated, entertained, and celebrated within the system". To Oviasuyi et al. (2010), "Corruption has become pervading, unabashed, uncontrolled and persistent…."

Quoting the Chairman of one of the anti-corruption agencies in Nigeria – the Independent Corrupt Practices and Other Related Offences Commission (ICPC), Fajonyomi & *Olu-Owolabi* (2013) pointed out that the local government has become a "bastion of corruption" because "more than 12 out of 54 criminal cases it was having in courts were against local government chairmen and its officials".

As Oruonye (2013) also puts it, corruption is so entrenched in the Nigerian local government system "so much so that practically nothing happens at the local government level". This may be a pointer to the notion that the local government may be the most corrupt tier of government in the country. This must have also accounted for why a former chairperson of the Economic and Financial Crimes and Other Related Offences Commission, Mrs. Farida Waziri, painted a gloomy picture of the situation when she posited:

... The waste of government resources at the council level had reached monumental proportions. The local government council in the country could not explain the mismanagement of over N3.313 trillion (N165/$) allocated to them across the country. Unfortunately, local government officials have not left their hands unsoiled in this regard... they drive to the council headquarters in their jeeps from the state capitals or the Federal Capital Territory, pay salaries and share other monies and disappear until it is time to share the next subvention. (Onwuemenyi, in Adeyemi, 2012)

To say the least, public service delivery in the local government of Nigeria has become so abysmal. This is where the application of E-governance comes in. The adoption of E-governance (ICT), cannot be overlooked in ensuring the improvement of public service delivery in the local government system of Nigeria. This study is therefore geared towards establishing the extent of ICT applications in the Local government and determine the problems therein to make its application result oriented.

REVIEW OF RELATED LITERATURE

The Concepts of E-Governance

E-governance has become a buzzword in public administration in this 21st century (Krishnan, 2013). It presents enormous opportunity to move governance forward by delivering qualitative, cost-effective public services to the citizenry. When e-governance is in place in any public sector organization, there is the prospect of government, having improved relationship with the general public, and its client (Nchuchuwe & Ojo, 2015).

E-governance is not just about institutions with computers and the Internet; its end result is to ensure that public services are provided to the general public in an efficient and effective manner that guarantees satisfaction. Thus as a public sector reform, e-governance is an inevitable tool for improving upon public service delivery (Nchuchuwe & Ojo, 2015). Ojo (2014) defines e-governance as the "...application of information communication technology by the government to enhance accountability, create awareness and ensure transparency in the management of government business." According to him, "it is a political strategy through which the activities of government are made known through the adoption of modern communication".

In the same vein, Estevez & Janowski (2013) conceives "Electronic Governance as the application of technology by government to transform itself and its interactions with customers, in order to create an impact on the society".

As Graham & Aurigi, (1997) put it, "many government agencies in developed countries have taken progressive steps toward the web and ICT use, adding coherence to all local activities on the Internet, widening local access and skills, opening up interactive services for local debates, and increasing the participation of citizens on promotion and management of the territory". The initiatives of government agencies and departments to use ICT tools and applications, Internet and mobile devices to support good governance, strengthen existing relationships and build new partnerships within civil society have become relevant in modern times. As with e-commerce, e-governance represents the introduction of a great wave of technological innovation as well as government reinvention. It represents a tremendous impetus to move forward in the 21st century with higher quality, cost effective government services and a better relationship between citizens and government (Fang, 2002).

Objectives of E-Governance

From the foregoing, it is glaring that e-governance aims to bridge the communication gap that had hitherto existed between the government and its customers (the citizens, businesses, employees and other government agencies). Nkwe (2012) outlines the objectives of e-governance to include the following:

- Cost reduction and efficiency gains;
- Quality of service delivery to businesses and customers;
- Transparency, anti - corruption, accountability;
- Increase in the capacity of government;
- Network and community creation;
- Improvement in the quality of decision – making;
- Promotion of the use of ICT in other sectors of the society.

Models of E-Governance

The models of E-governance have been identified by various authors to include four (4) broad categories. They are as follows:

- **Government to Citizens (G2C):** This model of e-governance is geared towards creating links and communication channels between the government and the citizens. The essence is to establish an efficient, free flow of interaction between the government and the citizens (Nkwe, 2012).
- **Government to Government (G2G):** This model aims at making ICT the focal point of inter-governmental relations (Gant, 2008). The G2G model is said to be the fulcrum of e-governance.

- **Government to Employees (G2E):** This model seeks to make better the interactions between the government and its employees through the application of ICTs.
- **Government to Business (G2B):** This model seeks to make transactions between the government and the private sector better via the use of ICT. The G2B type of e-governance is the linkage between the public and private sectors through the medium of the Internet, the web and mobile communications technology. (Gant, 2008).

At the local government level, all these models are applicable. However, this study is more concerned with the G2C and G2E models.

E-Governance Implementation in the Nigerian Public Service

The imposition of colonial rule by Britain in the geographical location now known as Nigeria in 1861 led to the establishment of structures and institutions of modern governance. Thus, the Nigerian public service is as old as colonial administration in Nigeria. Ayodele and Bolaji (2007) opined that "between 1900 and 1950, the Nigerian public service was dominated and controlled by British technocrats and generalist administrators". From the pre-independence to independence and post-independence eras, the Nigerian public service has undergone various reforms and transformations which were geared towards making it more efficient and effective or bring "improvement in the ways and manners in which government is managed and public goods and services are delivered effectively, and efficiently" (Osawe, 2014).

Though e-governance implementation in the Nigeria public service varies from level to level of government as well as agency to agency, there was actually an attempt at providing a unified, national framework of ICT adoption in governance. According to Olatokun and Adebayo (2012), the federal government of Nigeria, in 2001, announced ICT as a policy of national importance. This further culminated in the creation of a policy on information technology in the same year.

The intention by the Nigerian government to establish an ICT policy was predicated by the United Nation's adoption of the eight Millennium Development Goals in 2000, which set targets to reduce poverty, and ensure basic needs such as food, health, water, education, e.t.c., as well as guaranteeing the new benefits of new technologies, especially ICTs through public-private partnership (Olise, 2010).

The implementation of e-governance in Nigeria came with the realization that no country or its government can function properly in the information or digital age without the use of the web and the other mobile Internet technologies. With the need to reform the public service and make it ICT-compliant, Nigerian Federal

Government deemed it necessary to institute a national policy on ICT. Hence, in 2007, the enabling Act, the National Information Technology Development Act, was enacted by the National Assembly with an Agency established along with the Act, which was empowered to "plan, develop and promote the use of information technology in Nigeria" (Olatokun and Adebayo, 2012).

The National Information Technology Development Act of 2007 empowered various government agencies to embark on the implementation of ICTs in the public service. However, in 2011, the Federal Government created a new Ministry called the Ministry of Communications Technology, which was saddled with the mandate of streamlining ICT development and progress in line with the nation's plan for e-governance (Omerie and Omeire, 2014).

The National ICT draft policy was presented by the ministerial committee on ICT policy harmonization in 2012, which included several policy recommendations and reports. The following are the policy thrust of the National ICT draft policy of 2012, as it pertains to e-governance:

- To facilitate the implementation of e-government initiatives;
- To develop frameworks and guidelines, including interoperability and e-government framework for the enhanced development and use of ICT in the government;
- To develop and implement ICT training programs for public sector employees, in connection with introduction of e-government and other digital functions within government offices;
- To coordinate the integration of national e-government network infrastructure and services; and
- To promote e–government and other e-services that would foster broadband usages (National ICT Policy, 2012).

The private sector was also included in the process of implementing e-governance in the Nigerian public service. The National Information Technology Development Act of 2007 made provisions for the avenue of public-private partnership in the adoption and management of ICTs in Nigeria by establishing the NeGST- National e-Government Strategies limited.

The NeGST was structured to be a three-party joint venture with shared roles and responsibilities. It comprised of the government (which is represented by NITDA), private and financial investors, and technology partners, with 5%, 15% and 80% share ownership in the joint venture, respectively (Omeire & Omeire, 2014). The essence of this strategic tripartite alliance called the NeGST was to create a viable unified national framework for the adoption and implementation of ICTs in the public service and their customers. The website of NeGST succinctly states that the purpose of its creation is to facilitate drive, and implement the Nigerian e-government programme under a public-private partnership model (Mundy & Musa, 2010).

PROSPECTS OF E-GOVERNANCE IMPLEMENTATION FOR PUBLIC SERVICE DELIVERY IN NIGERIA

Service delivery is more complex in the public sector as it is not just about meeting expressed needs, but finding out the needs that are not expressed, setting priorities, resource allocation and publicly justifying and accounting for what has been done (Darma and Ali, 2014). For developing countries like Nigeria, the public sector plays major tasks in service delivery since the government is both employer of labour and owner of resources. In view of this, public agencies have come under pressure and debate in recent years in terms of their efficiency and effectiveness in service delivery with respect to the needs and rights of the citizens.

The Nigerian public service is a stabilizing force for national unity and the core of government activities. It guarantees national cohesion and continuity in governance of the country. The crisis of service delivery and the perennial complaints of inefficiency and ineffectiveness on the Nigerian public service over the decades in Nigeria could be addressed by the immense benefits and prospects of e-governance implementation in the Nigerian bureaucracy (Nchuchuwe & Ojo, 2015).

The position of e-governance in Nigerian public service holds enormous advantages that can ensure improvements in service delivery. According to Coleman (2005), the prospects of e-governance implementation ensures that:

- A freer flow of information between departments, agencies and layers within government;
- More professional administrators, supported by standardized, electronically-embedded decision-making system;
- The routine provision of service according to impersonal rules, as opposed to clients arrangements;
- Transparency, particularly in relations to the procurement of government services;
- Opportunities to work in partnership with the private sector in modernizing government processes;
- A freer flow of information between government and citizens;
- The strengthening of intermediary democratic institutions, such as parliaments, local government, civil society organizations and independent media;

- Opportunities for citizens to participate more directly in policy development;
- Opportunities to combine traditional and modern methods of accountability.

METHODOLOGY

The study adopted the survey method and other secondary sources to gather data. The population of the study is the employees of Ojo and Alimosho local government areas totaling 748, while the sample size for the two local government areas is 100, with 50 respondents selected from each of the local government areas using the simple random technique.

Structured questionnaires were designed to elicit responses from the respondents. The questionnaire was administered by the researchers in person.

The following hypotheses were formulated for the study:

Hypothesis One

H_0: Local government in Nigeria does not have all the facilities (web sites, electricity supply, computer/internet, e-centers etc) for E-governance application.

H_1: Local government in Nigeria has all the facilities (web sites, electricity supply, computer/internet, e-centers etc) for E- governance application.

Hypothesis Two

H_0: E-governance has not enhanced service delivery in the local government of Nigeria.

H_1: E-governance has enhanced service delivery in the local government of Nigeria.

Hypothesis Three

H_0: E-governance has not engendered the participation of the local communities in decision-making.

H_1: E-governance has engendered participation of the local communities in decision-making.

Hypothesis Four

H_0: E-governance has not engendered transparency and accountability in the services of the local government.

H_1: E-governance has engendered transparency and accountability in the services of the local government.

Hypothesis Five

H_0: E-governance has not engendered improved internally generated revenue in the services of the local government.
H_1: E-governance has engendered improved internally generated revenue in the services of the local government.

The data from the questionnaire was analyzed section by section, using bar charts, tables, and simple percentages.

50 questionnaires were distributed to each of the local government area. At the end, 47 questionnaires were retrieved from Ojo local government area, while 49 questionnaires were retrieved from Alimosho local government area.

Classification of Respondents' Characteristics

This section covers the biographical data of respondents, such as sex, age distribution, marital status, educational qualification, working experience, e.t.c.

Table 1 reveals that the respondents were made up of more men than women. This is not unexpected as more men dominate the public service in Nigeria especially the local government. Below is a bar chart presentation of respondents by sex distribution for the two local government areas:

Table 2 reveals that there are more married respondents in Ojo local government than in Alimosho local government which has more unmarried respondents than other categories. This shows that there is no marital status bias in the local government areas under study. Figure 2 shows a bar chart presentation of respondents by marital status for the two local government areas.

Table 3 reveals that the two local governments have high number of respondents within the age range of 36 – 45 years. This signifies that most of the respondents are

Table 1. Sex distribution of respondents

Local Govt	Sex Distribution		Total
	Male	Female	
Ojo	26	21	47
Alimosho	33	16	49

Source: Field study, March, 2016.

Figure 1. Respondents by sex distribution
Source: Field study, March, 2016.

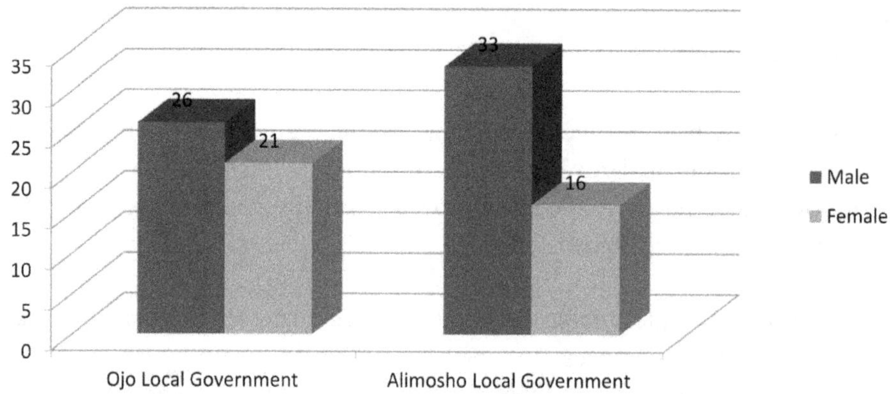

Table 2. Marital status of respondents

Local Govt	Marital Status			Total
	Single	Married	Others	
Ojo	13	18	16	47
Alimosho	23	19	07	49

Source: Field study, March, 2016.

Figure 2.
Source: Field study, March, 2016.

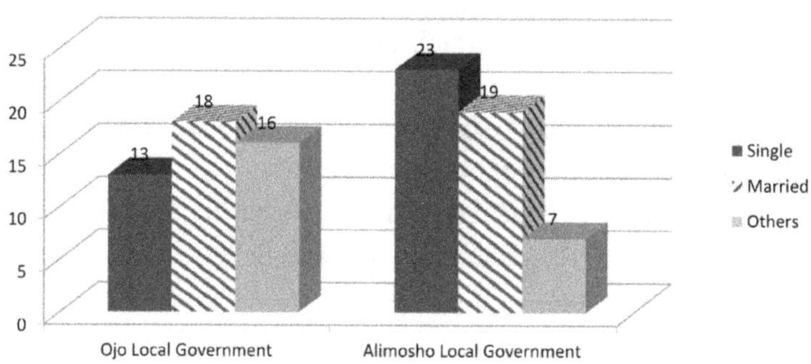

matured enough to respond meaningfully to the issues raised in the study. Figure 3 shows a bar chart presentation of respondents by age distribution for the two local government areas.

Table 4, reveals that majority of the respondents have good education with many having secondary school leavers' certificates. This shows that they are knowledgeable

Table 3. Age distribution of respondents

Local Govt	Age Distribution (in Years)					Total
	Below 25	25 - 35	36 - 45	46 - 55	Above 55	
Ojo	03	09	17	14	4	47
Alimosho	04	17	21	06	1	49

Source: Field study, March, 2016.

Figure 3. Respondents by age distribution
Source: Field study, March, 2016.

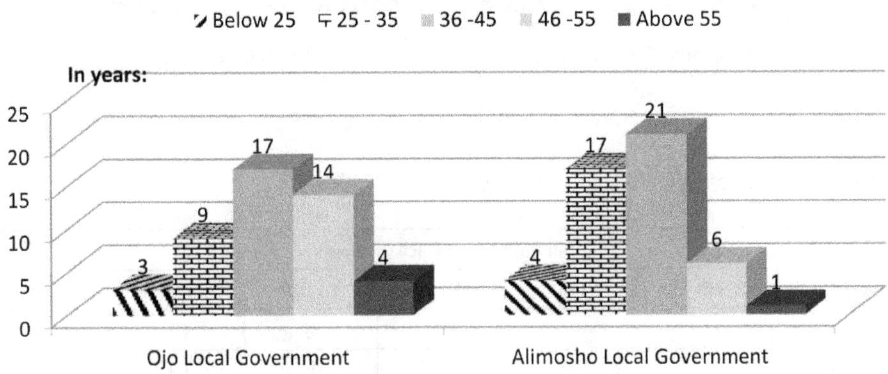

Table 4. Level of educational qualification of respondents

Local Govt	Age Distribution (in Years)					Total
	No Formal Education	Primary Education	Secondary Education	Tertiary Education	Above First Degree	
Ojo	10	08	13	05	11	47
Alimosho	07	05	22	06	09	49

Source: Field study, March, 2016.

enough to respond meaningfully to the Questions raised. Figure 4 shows a bar chart presentation of respondents by level of educational qualification for the two local government areas:

In table 5, most of the employees in Ojo local government have working experience of between 6 to 10 years, while in Alimosho local government most of the employees have working experience of between 11 to 20 years. This shows that majority of the respondents have long working experience to ascertain the effect of E-governance

Figure 4. Respondents by level of educational qualification
Source: Field study, March, 2016.

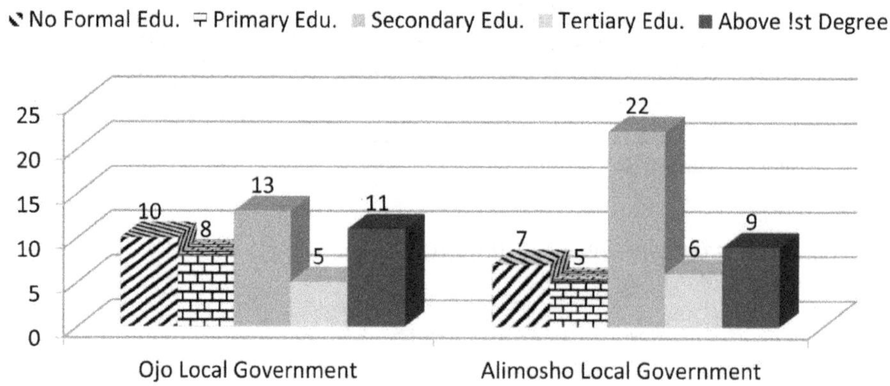

Table 5. Working experience of respondents

Local Govt	Working Experience (in Years)					Total
	Below 5	6 -10	11 - 15	16 - 20	Above 20	
Ojo	11	17	03	09	07	47
Alimosho	07	09	23	05	03	49

Source: Field study, March, 2016.

Figure 5. Respondents by working experience
Source: Field study, March, 2016.

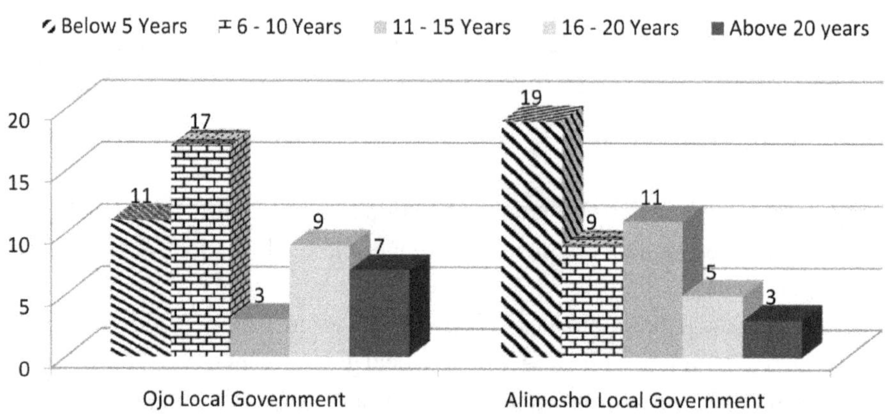

Table 6. Staff categories of respondents

Local Govt	Staff Categories			Total
	Junior	Senior	Management	
Ojo	09	32	06	47
Alimosho	07	39	03	49

Source: Field study, March, 2016.

Table 7. Religions of respondents

Local Govt	Religion			Total
	Christianity	Islam	Others	
Ojo	25	15	07	47
Alimosho	18	23	06	49

Source: Field study, March, 2016.

Table 8. Respondents on whether the use of ICT (Information and Communications Technology) has improved service delivery in Ojo and Alimosho local governments

Variables	Local Government			
	Ojo		Alimosho	
	Frequency	Percentage	Frequency	Percentage
Strongly Disagreed	18	38.3	13	26.5
Disagreed	17	36.2	29	59.2
Undecided	02	4.3	01	2.0
Agreed	07	14.9	04	8.2
Strongly Agreed	03	6.4	02	4.1
Total	47	100	49	100

Source: Field study, March, 2016.

on service delivery in the local governments. Figure 5 shows a bar chart presentation of respondents working experience for the two local government areas.

Table 6 reveals that the two local governments have the highest number of respondents in the senior staff category. This shows that majority of the respondents are in better official positions to know what obtains in the local government.

Table 7, reveals that most of the respondents in Ojo local government are Christians, while in Alimosho local government, most of the respondents are Muslims. This shows that the study is not religiously biased.

PRESENTATION AND ANALYSIS OF RESEARCH STATEMENTS

Under this section, the responses from the respondents were analyzed and presented using Descriptive statistics namely, tables and simple percentages.

Table 8, shows that 38.3% and 36.2% of respondents from Ojo local government strongly disagreed and disagreed respectively that the use of ICT has improved service delivery in the local government; 14.9% and 6.4% agreed and strongly agreed respectively that the use of ICT has improved service delivery, while 4.3% were undecided. On the other hand, at Alimosho local government 26.5% strongly disagreed, 59.2% disagreed, 2.0% were undecided, 8.2% agreed and 4.1% strongly agreed. This implies that majority of the respondents are of the view that the use of ICT has not improved service delivery in these local governments.

Table 9 reveals that 36.2% and 34.0% of the respondents from Ojo local government strongly disagreed and disagreed respectively that ICT has brought close relationship between the staff and the public in the local governments; 6.4% and 14.9% agreed and strongly agreed respectively, while 8.5% were undecided. On the other hand, in Alimosho local government, 32.7% strongly disagreed, 34.7% disagreed, 4.1% were undecided, 18.4% agreed and 10.27% strongly agreed. This implies that workers in the two local governments disagreed that ICT has brought close relationship between government staff and the public

Table 10 reveals that 23.4% and 38.3% of respondents from Ojo local government strongly disagreed and disagreed respectively that the public contribute to policy and decision-making through ICT; 19.1% and 14.9% agreed and strongly agreed respectively, while 4.3% were undecided. On the other hand, in Alimosho local government, 26.5% strongly disagreed and 30.6% disagreed while 20.4% agreed,

Table 9. Respondents on whether ICT has promoted close relationship between government staff and the public in Ojo and Alimosho local governments

	Local Government			
	Ojo		Alimosho	
Variables	Frequency	Percentage	Frequency	Percentage
Strongly Disagreed	17	36.2	16	32.7
Disagreed	16	34.0	17	34.7
Undecided	03	6.4	02	4.1
Agreed	07	14.9	09	18.4
Strongly Agreed	04	8.5	05	10.2
Total	47	100	49	100

Source: Field study, March, 2016.

Table 10. Responses on whether the public contributes to policy and decision-making through ICT in Ojo and Alimosho local governments

	Local Government			
	Ojo		Alimosho	
Variables	Frequency	Percentage	Frequency	Percentage
Strongly Disagreed	11	23.4	13	26.5
Disagreed	18	38.3	15	30.6
Undecided	02	4.3	03	6.1
Agreed	09	19.1	10	20.4
Strongly Agreed	07	14.9	09	18.4
Total	47	100	49	100

Source: Field study, March, 2016.

Table 11. Respondents on whether ICT has made work faster, interesting and less cumbersome in Ojo and Alimosho local governments

	Local Government			
	Ojo		Alimosho	
Variables	Frequency	Percentage	Frequency	Percentage
Strongly Disagreed	03	6.4	05	10.2
Disagreed	07	14.9	13	26.5
Undecided	01	2.1	06	12.2
Agreed	16	34.0	16	32.7
Strongly Agreed	10	21.3	09	18.4
Total	47	100	49	100

Source: Field study, March, 2016.

18.4% strongly agreed, 6.1% were undecided. From the analyses, it is clear that policy making is still centered on the political elites in the local governments.

Table 11 reveals that 6.4% and 14.9% of respondents from Ojo local government strongly disagreed and disagreed respectively that ICT has made work faster, interesting and less cumbersome; 34.0% and 21.3% agreed and strongly agreed respectively while 2.1% were undecided. Similarly, in Alimosho local government, 10.2% strongly disagreed, 26.5% disagreed, 12.2% were undecided, 32.7% agreed and 18.4% strongly agreed. This shows that majority of the respondents are of the view that ICT has made work faster, interesting and less cumbersome in the Nigerian local government system.

Table 12. Respondents on whether financial wastages and leakages have not been reduced since the use of ICT in Ojo and Alimosho local governments

	Local Government			
	Ojo		Alimosho	
Variables	Frequency	Percentage	Frequency	Percentage
Strongly Disagreed	06	12.8	07	14.3
Disagreed	07	14.9	01	2.0
Undecided	05	10.6	02	4.1
Agreed	18	38.3	18	36.7
Strongly Agreed	11	23.4	21	42.9
Total	47	100	49	100

Source: Field study, March, 2016.

Table 13. Respondents on whether the needs of citizens are quickly recognized and treated through the use of ICT in Ojo and Alimosho local governments

	Local Government			
	Ojo		Alimosho	
Variables	Frequency	Percentage	Frequency	Percentage
Strongly Disagreed	07	14.9	07	14.3
Disagreed	03	6.4	10	20.4
Undecided	01	2.1	01	2.0
Agreed	17	36.2	16	32.7
Strongly Agreed	19	40.4	15	30.6
Total	47	100	49	100

Source: Field study, March, 2016.

Table 12 reveals that 12.8% and 14.9% of respondents from Ojo local government strongly disagreed and disagreed respectively that financial wastage and leakages have been reduced since the use of ICT in the local government; 38.3% and 23.4% agreed and strongly agreed respectively while 10.6% were undecided. Similarly, in Alimosho local government, 14.3% strongly disagreed, 2.0% disagreed, 4.1% were undecided, 36.7% agreed and 42.9% strongly agreed. This implies that the respondents were favourably disposed to the view that wastages and leakages have not been reduced on the use of ICT in the local governments. This is also a pointer that E-governance is not fully applied in the local governments.

Table 14. Respondents on whether there are no complaints from the public as a result of the use of ICT for service delivery in Ojo and Alimosho local governments

	Local Government			
	Ojo		Alimosho	
Variables	**Frequency**	**Percentage**	**Frequency**	**Percentage**
Strongly Disagreed	16	34.0	13	26.5
Disagreed	17	36.2	11	22.4
Undecided	03	6.4	04	8.2
Agreed	04	8.5	17	34.7
Strongly Agreed	07	14.9	04	8.2
Total	47	100	49	100

Source: Field study, March, 2016.

Table 15. Respondents on whether ICT has ensured an efficient and effective financial management system in Ojo and Alimosho local governments

	Local Government			
	Ojo		Alimosho	
Variables	**Frequency**	**Percentage**	**Frequency**	**Percentage**
Strongly Disagreed	06	12.8	07	14.3
Disagreed	07	14.9	04	8.2
Undecided	05	10.6	05	10.2
Agreed	18	38.3	17	34.7
Strongly Agreed	11	23.4	16	32.7
Total	47	100	49	100

Source: Field study, March, 2016.

Table 13 reveals that 14.9% and 6.4% of respondents from Ojo local government strongly disagreed and disagreed respectively that the needs of the citizens are quickly recognized and treated through the use of ICT; 36.2% and 40.4% agreed and strongly agreed respectively. On the other hand, in Alimosho local government, 14.3% strongly disagreed, 20.4% disagreed, 2.0% were undecided, 32.7% agreed and 30.6% strongly agreed. This implies that the respondents appreciate that the needs of local citizens can quickly be recognized and treated through the use of ICT in the local governments.

Table 14 shows that 34.0% and 36.2% of the respondents from Ojo local government strongly disagreed and disagreed respectively that there are no complaints from

Table 16. Respondents on whether revenue generation has not greatly improved in Ojo and Alimosho local governments with ICT applications

Variables	Local Government			
	Ojo		Alimosho	
	Frequency	Percentage	Frequency	Percentage
Strongly Disagreed	10	21.3	11	22.4
Disagreed	05	10.6	09	18.4
Undecided	03	6.4	01	2.0
Agreed	16	34.0	13	26.5
Strongly Agreed	13	27.7	15	30.6
Total	47	100	49	100

Source: Field study, March, 2016.

the public because of the use of ICT in service delivery; 8.5% and 14.9% agreed and strongly agreed respectively while 6.4% were undecided. On the other hand, in Alimosho local government, 26.5% strongly disagreed, 22.4% disagreed, 8.2% were undecided, 34.7% agreed and 8.2% strongly agreed. This implies that majority of the respondents share similar views that the use of ICT in the two local governments for service delivery in Nigeria has led to some complaints from the public.

Table 15 shows that 12.8% and 14.9% of the respondents from Ojo local government strongly disagreed and disagreed respectively that ICT has ensured an efficient and effective financial management system; 38.3% and 23.4% agreed and strongly agreed respectively while 10.6% were undecided. On the other hand, in Alimosho local government, 14.3% strongly disagreed, 8.2% disagreed, 10.2% were undecided, 34.7% agreed and 32.7% strongly agreed. This implies that majority of the respondents share the view that ICT has engendered an efficient and effective financial management system in Nigerian local governments.

Table 16 shows that 21.3% and 10.6% of the respondents from Ojo local government strongly disagreed and disagreed respectively that revenue generation has not greatly improved; 34.0% and 27.7% agreed and strongly agreed respectively while 6.4% were undecided. On the other hand, in Alimosho local government, 22.4% strongly disagreed, 18.4% disagreed, 2.0% were undecided, 26.5% agreed and 30.6% strongly agreed. This implies that majority of the respondents' share the view that revenue generation has not greatly improved in the two local governments.

Table 17 shows that 19.1% and 10.6% of the respondents from Ojo local government strongly disagreed and disagreed respectively that workers derive greater job satisfaction via the use of ICT in their jobs; 40.4% and 27.7% agreed and strongly agreed respectively while 2.1% were undecided. On the other hand, in

Table 17. Respondents on whether workers derive greater job satisfaction via the use of ICT in their jobs in Ojo and Alimosho local governments

	Local Government			
	Ojo		Alimosho	
Variables	Frequency	Percentage	Frequency	Percentage
Strongly Disagreed	09	19.1	06	12.2
Disagreed	05	10.6	09	18.4
Undecided	01	2.1	04	8.2
Agreed	19	40.4	18	36.7
Strongly Agreed	13	27.7	12	24.5
Total	47	100	49	100

Source: Field study, March, 2016.

Table 18. Respondents on whether transparency and accountability have not improved since the adoption of ICT in Ojo and Alimosho local governments

	Local Government			
	Ojo		Alimosho	
Variables	Frequency	Percentage	Frequency	Percentage
Strongly Disagreed	19	40.4	16	32.7
Disagreed	14	29.8	13	26.5
Undecided	05	10.6	07	14.3
Agreed	06	12.8	09	18.4
Strongly Agreed	03	6.4	04	8.2
Total	47	100	49	100

Source: Field study, March, 2016.

Alimosho local government, 12.2% strongly disagreed, 18.4% disagreed, 8.2% were undecided, 36.7% agreed and 24.5% strongly agreed. This implies that majority of the respondents are of the view that workers derive greater job satisfaction via the use of ICT in their jobs in the two local governments.

Table 18 shows that 40.4% and 29.8% of respondents from Ojo local government strongly disagreed and disagreed respectively that transparency and accountability have not improved since the adoption of ICT in the two local governments; 12.8% and 6.4% agreed and strongly agreed respectively while 10.6% were undecided. On the other hand, in Alimosho local government, 32, 7% strongly disagreed, 26.5% disagreed, 14.3% were undecided, 18.4% agreed and 8.2% strongly agreed. With

Table 19. Respondents on whether ICT will modernize Ojo and Alimosho local governments, and make them more dynamic in their activities

Variables	Local Government			
	Ojo		Alimosho	
	Frequency	Percentage	Frequency	Percentage
Strongly Disagreed	17	36.2	13	26.5
Disagreed	14	29.8	19	38.8
Undecided	02	4.3	04	8.2
Agreed	09	19.1	05	10.2
Strongly Agreed	05	10.6	08	16.3
Total	47	100	49	100

Source: Field study, March, 2016.

Table 20. Respondents on whether the two Local governments have all the facilities (web sites, electricity supply, computer/internet, e-centers etc) for E- governance application

Variables	Local Government			
	Ojo		Alimosho	
	Frequency	Percentage	Frequency	Percentage
Strongly Disagreed	12	25.53	11	22.45
Disagreed	30	63.83	34	69.39
Undecided	0	0	1	2.04
Agreed	4	8.51	3	6.12
Strongly Agreed	1	2.13	0	0
Total	47	100	49	100

Source: Field study, March, 2016.

majority of the respondents disagreeing, it shows that transparency and accountability have improved in the two local governments since the adoption of ICT.

Table 19 shows that 36.2% and 29.8% of the respondents from Ojo local government strongly disagreed and disagreed respectively that ICT will not modernize and make the local government system more dynamic in its activities; 19.1% and 10.6% agreed and strongly agreed respectively while 4.3% were undecided. On the other hand, in Alimosho local government, 26.5% strongly disagreed, 38.8% disagreed, 8.2% were undecided, 10.2% agreed and 16.3% strongly agreed. This implies that majority of the respondents share the view that ICT will bring modernization to the two local governments and will make them more dynamic in the use of modern technology for their activities.

Table 20 shows that 25.5% and 63.8% of the respondents from Ojo local government strongly disagreed and disagreed respectively that the Local government has all the facilities (web sites, electricity supply, computer/internet, e-centers etc) for E- governance application while non was indifferent. 8.5% and 2.13% strongly agreed and agreed respectively. Similarly, in Alimosho local government, 22.4% and 69.3% strongly disagreed and disagreed respectively with only one respondent indifferent, while 6.12% only agreed. None disagreed. The implication of this is that the respondents are of the view that their local governments are not e-ready.

Test of Hypotheses

The following formulated hypotheses for the study were tested using descriptive statistics.

Hypothesis One

H_0: Local government in Nigeria does not have all the facilities (web sites, electricity supply, computer/internet, e-centers etc) for E-governance application

H_1: Local government in Nigeria has all the facilities (web sites, electricity supply, computer/internet, e-centers etc) for E- governance application

Tables 13 and 20 were used to test this hypothesis. From table 13, the respondents agree that service provisions to the people have not been through the application of ICT. 69.8% held this view. Similarly, it was clear from the responses on table 20 that the two local governments under study are far from being E-governance ready. This is because the necessary facilities required for the application are not present. For E-governance to take off and be successful such facilities like electricity, ICT trained personnel, web sites, computers etc must be available. Since majority of the respondents (90.63%) disagreed that the facilities are available, the null hypothesis is accepted while the alternative hypotheses is rejected.

Hypothesis Two

H_0: E-governance has not enhanced service delivery in the local government of Nigeria.

H_1: E-governance has enhanced service delivery in the local government of Nigeria.

Tables 8 and 11, were used to test this hypotheses. From the tables, it was clear that service delivery in the two local governments has not been enhanced. The implication of this is that E-governance has not been put into full application in the two local governments. From table 8, majority of the respondents (80.21%) concurred that service delivery is still very poor. From table 11, the respondents concurred that service delivery has not been made faster, interesting and less cumbersome in the two local governments.53.2% held this view in the two local governments. For this reason the null hypothesis is accepted and the alternative hypothesis is rejected. The point is clear that with E-governance service delivery can be enhanced and become faster and cumbersome.

Hypothesis Three

H_0: E-governance has not engendered the participation of the local communities in decision-making.

H_1: E-governance has engendered participation of the local communities in decision-making.

Table 10 was used to test this hypothesis. The respondents (59.37%) disagreed with the suggestion that the citizens make contributions to policy making in the two local governments, Contributions to decision making is one of the hallmarks of E-governance. The implication of this is that the leaders of the local government are still elitist hovering the policy making process among themselves. The null hypothesis is thus accepted and the alternative hypothesis rejected.

Hypothesis Four

H_0: E-governance has not engendered transparency and accountability in the services of the local government.

H_1: E-governance has engendered transparency and accountability in the services of the local government.

Table 18 was used to test this hypothesis. The respondents in the two local governments concur that Accountancy and Transparency have not improved in the local governments as majority of the respondents (69.2%) disagreed. Again accountability and transparency are hallmarks of E- governance and its non-improvement in the local governments cannot be completely separated from the fact that there is poor or non application of E-governance in the local governments. The null hypothesis is therefore accepted and the alternative hypothesis rejected.

Hypothesis Five

H_0: E-governance has not engendered improved internally generated revenue in the services of the local government.

H_1: E-governance has engendered improved internally generated revenue in the services of the local government.

Tables 12 and16 were used to test the hypothesis. Majority of the respondents (70.8%) in table 12 disagreed that financial wastages and leakages have reduced in the local governments. Ditto for table 16 where the respondents (59.4%) jointly agreed that revenue generation is still poor in the local governments. The null hypothesis is thus accepted and the alternative hypothesis rejected.

DISCUSSIONS OF MAJOR FINDINGS

Having analyzed and interpreted the data obtained from the two local governments (Ojo and Alimosho) coupled with the various interviews the researchers had with officials of the local governments, it became clear that the application of E-governance is still a far cry in the local government areas. The researchers observed that the concept was strange to many of the officials of the local governments. Only few knew what it was while many interpreted it to mean using electronic means to pay staff salaries and direct lodgments into the bank by tax payers and other who are in transaction with the local governments. Throughout the period of the study, the researchers did have the opportunity of meeting with the chairmen of the local governments but some officials who sought for protection revealed that the chairmen hardly come or stay in the office. This is in line with the views expressed by a former chairperson of EFCC, Mrs Farida Waziri that "Local government has become so far removed from the lives of the people to a point where some Chief Executives of local council no longer reside in the domains they were elected to administer. They drive to the council headquarters in their jeeps from the state capitals or the Federal Capital Territory, pay salaries and share other monies and disappear until it is time to share the next subvention" (Onwuemenyi, in Adeyemi, 2012). The application of E-governance requires the full attention of the leaders in the local governments and where this is not present adoption and implementation of a project like E-governance will become difficult.

It was also observed that the local governments were generally short of been ready for E-governance. None of the major facilities required for E-governance was available. Though there was a claim by some of the officials that there exist websites in the local governments but there were largely ineffective and available for public use. E-governance can only operate were websites are available for various information and access. Similarly, though some of the officials claim to have worked and acquired skills working with the local government, they have never been trained for skills on effective E-governance implementation. This also goes to buttress the point that the local governments are truly not e-ready. Also, there is the problem of poor electricity supply and unavailability of computers or other forms of telecommunications. The researchers also observed that many of the officials interacted with seem not to like the idea of E-governance for what some of them considered capable of sending them to the labour market. The implication of this is that the amount of cooperation a project like E-governance may require may become elusive thereby hindering its application. When the researchers interviewed some of the local citizens it was observed that majority of them lack knowledge of what E-governance is about. Aside from this they also complained about electricity supply, lack of computer knowledge, and non-possession of computers or laptops-useful instruments for E-governance applications. Having said these, efforts have to be made to ensure that the wherewithal required for E-governance to be applied in the local governments should be put in place if the local governments must strive towards effective service delivery in the 21st century.

CONCLUSION

E-government is a concept that has been defined from various perspectives by various authors. However, the core of e-governance is on the provision of governmental services via the use of information and communication technology. Generally, the success of e-governance is geared towards the improvement of services to the citizens among others in a faster, cheaper and effective manner. The importance of e-governance in public service delivery cannot be over emphasized.

As Fang (2002), puts it "E-Governance represents the introduction of a great wave of technological innovation as well as government reinvention. It represents a tremendous impetus to move forward in the 21st century with higher quality, cost effective government services and a better relationship between citizens and government".

With e-governance, the public can expect speedy, enhanced, affordable, and more accessible services from the government. Also, e-governance tends to build bridges between the government and the governed, by creating better relationships, more transparency, more engagement and trust.

This study critically examined the state of e-governance application in Ojo and Alimosho local governments of Nigeria as it relates to their services to the public. The study was able to ascertain the possibility of the Nigerian local governments to practice e-governance with the aim of improving service delivery, reducing wastages and other factors that militate against efficiency and effectiveness. However, there are a lot of gaps, which were revealed from the study which must be filled for E-governance to be applied.

RECOMMENDATIONS

This study recommends a re-visit of the country's ICT policy, especially with regards to e-governance implementation in the local governments. There should be a concerted effort towards providing adequate communication facilities towards a complete computerization of local government operations, and transactions with the public.

Moreover, the study advocates that practical steps be taken with the purpose of ensuring that there is sufficient and stable supply of electricity, public enlightenment and the development of human resource so as to accomplish top computer literacy level and ICT compliance among the vast majority of the local public and employees.

Also, the accessibility of internet facilities and services must increase among the populace by ensuring that it is less expensive, to all Nigerians, regardless of which ever part of the country they reside. The world is on the move and Nigeria is an integral part of it, hence, should not lag behind in terms of the use of ICT.

In addition, the Nigerian government must come to the awareness that the world as it is now is completely digitalized, and that no organization can efficiently overcome the realities and challenges of this modern times without integrating its system into the world of ICT. Thus, the public sector cannot continue to do things the old ways it has been used to doing over the years and expect different results. To achieve the much desired results, there must be a paradigm shift from the old methods of getting things done, by inculcating and imbibing new ideas that will make work faster, less cumbersome and at the same time engender development in vital sectors of the economy. The implementation of e-governance in Nigeria's local government and the public sector in general is, certainly, an innovative and far-reaching reform, which would invariably result in the improvement of public service delivery.

Hence, it behooves the Nigerian government to make concerted efforts to ensure that the implementation of e-governance in the public sector is not only embraced, sustained, but becomes the driving force for efficient and effective service delivery to the general public. Any nation that is seeking to become a key player in the global scheme of things must not be left behind in this Information Age. In the world today, no country can thrive on and experience economic growth and prosperity on the basis of its natural resources without keying into ICT (e-governance). To successfully have

ICT running unhindered and fully functional in the public sector, deliberate plans should be made to ensure that mass education, enlightenment campaign on ICT and e-governance system, provision of enabling environment and power supply are available. The Federal government has the potential of putting Nigeria on the tract towards accomplishing efficient and effective public service delivery at all levels of government in the 21st century with E-governance if the political will is there.

REFERENCES

Adeyemi, O. O. (2012). Corruption and Local Government Administration in Nigeria: A Discourse of Core Issues. *European Journal of Sustainable Development.*, *1*(2), 183–198.

Arowolo, D. (2012). Ethics, motivation and performance in Nigeria's public service. *Public Policy and Administration Research*, *2*(5), 37–44.

Ayodele, B., & Bolaji, K. (2007). Public service and democracy in developing societies: The Nigerian experience. *Journal of Social Science*, *15*(2), 105–110.

Coleman, S. (2005). *African E-governance – opportunities and challenges*. Oxford Internet Institute, University of Oxford.

Darma, N. A., & Ali, M. (2014). An assessment of public sector service delivery in Nigeria: A case study of federal capital territory area councils, 2007-2011. *International Journal of Development and Sustainability.*, *3*(8), 1756–1776.

Edewor, P. A., Imhonopi, D., & Urim, U. M. (2014). ICTs and sustainable development of higher education in Nigeria: Rewriting the ugly narrative. *Journal of Educational and Social Research.*, *4*(1), 357–365.

Estevez, E., & Janowski, T. (2013). Electronic governance for sustainable development — conceptual framework and state of research. *Government Information Quarterly*, *30*(1), 94–109. doi:10.1016/j.giq.2012.11.001

Fajonyomi, B., & Olu-Owolabi, F. E. (2013, April). Good Governance and Local Government Administration for Development: Multiple Perspectives. *Journal of Capital Development in Behavioural Sciences*, *1*, 1–16.

Fang, Z. (2002). E-Government in Digital Era: Concept, Practice and Development. *International Journal of the Computer*, *10*(2), 1–22.

Gant, J. P. (2008). The electronic government for developing countries. Policies and Strategies, International Telecommunication Union, August 2008 DRAFT, Geneva.

Graham, S., & Aurigi, A. (1997). Virtual Cities, Social Polarisation, and the Crisis in Urban Public Space. *Journal of Urban Technology*, *4*(1), 19–52. doi:10.1080/10630739708724546

Krishman, C. (2013). E-*Governance in Local Governance: The case of Kerala.* Retrieved from www.iimahd.ernet.in/.../krishnan.htm

Mundy, D., & Musa, B. (2010). Towards a framework for e-government development in Nigeria. *Electronic. Journal of E-Government*, *8*(2), 148–161.

National ICT Policy. (2012). *National ICT Draft Policy*. Presented by the Ministerial Committee on ICT Policy Harmonisation.

Nchuchuwe, F. F., & Ojo, A. D. (2015). Challenges and Prospects of Implementing E-Governance in Nigeria. *Covenant University Journal of Politics and International Affairs*, *3*(2), 1–19.

Nkwe, N. (2012). E-government: Challenges and opportunities in Botswana. *International Journal of Humanities and Social Science*, *2*(17), 39–48.

Ojo, J. S. (2014). E-Governance: An imperative for sustainable grass root development in Nigeria. *Journal of Public Administration and Policy Research.*, *6*(4), 77–89. doi:10.5897/JPAPR2013.0264

Olatokun, W. M., & Adebayo, B. M. (2012). Assessing e-government implementation in Ekiti State, Nigeria. *Journal of Emerging Trends in Computing and Information Sciences.*, *3*(4), 497–505.

Olise, P. F. (2010). Information and communication technologies (icts) and sustainable development in Africa: Mainstreaming the Millennium Development Goals (MDGs) into Nigeria's development agenda. *Journal of Social Science.*, *24*(3), 155–167.

Omeire, E., & Omerie, C. (2014). New wine in old wine skin: An exploration of major constraints to e-government implementation in Nigeria. *European Scientific Journal*, *10*(14), 481–487.

Oruonye, E. D. (2013). Grassroot Democracy and the Challenges of Rural Development in Nigeria: A Case Study of Bali Local Government Area of Taraba State. *Asian Journal of Humanities and Social Sciences*, *1*(1), 12–19.

Osawe, C. O. (2014). Reengineering professionalism in the Nigerian public service towards service delivery. *Review of Public Administration and Management*, *3*(6), 63–76.

Oviasuyi, P. O., Idada, W., & Isiraojie, L. (2010). Constraints of Local Government Administration in Nigeria. *Journal of Social Sciences*, *24*(2), 81–86.

KEY TERMS AND DEFINITIONS

Application: Implementation or execution.

E-Governance Facilities: These are the facilities useful for the application of e-governance for example, web sites, constant electricity supply, electronic devices or computers, internet diffusions, people computer literacy etc.

E-Governance: A system of governance that is electronically applied in which human interactions are reduced for effective communications.

Interrogation: Investigating or probing into an event or situation to establish facts.

Local Government: Government at the local levels or grass root closer to the people. It can also be defined as the government of the people by the people for the people at the local levels.

Public Enlightenment: Creating awareness on the populace for the effective use of ICT for service delivery.

Service Delivery: Providing the basic necessities of life for the citizens.

APPENDIX: QUESTIONNAIRE FOR THE STUDY ADMINISTERED TO THE EMPLOYEES OF OJO AND ALIMOSHO LOCAL GOVERNMENT

Table 21.

		Yes	No	I Can't Say
1	Are you aware of the application of E-governance in your Local government?			
2	Have you been trained to use the website?			
3	Have you used the local government websites before?			
4	Is your local government website linked to other websites?			
5	Do you use the website to communicate with other staff?			
6	Is the Information on your local government website up to date?			
7	Does the citizen have access to Information in your local government by electronic means?			
8	Does the website allow you to relate with the public online?			
9	Have you ever rendered services online using the website?			
10	Is there a budget to manage your local government website?			
11	Does the website allow you to relate with political leaders online?			
12	Do you relate with other political leaders online?			
13	Using the website you can express your views			
14	Do you trust that information posted on the website is secure?			
15	Have you ever sought for people's views on the website?			
16	Have you received feedback using the website?			
17	Have you completely migrated to using the website?			
18	The website has made communication between staff and citizens easy			
19	It has reduced on daily working costs			
20	It has enabled faster decision making			
21	It has increased transparency in government			
22	It has reduced the level of corruption			

Table 22.

		Adequately Available	Available	Scantily Available	Not Available	I Cannot Say
1	Funds to sustain the website.					
2	IT skills among staff members					
3	Political will by politicians.					
4	Trust in the website due to past system failures.					
5	Availability of computers.					
6	Power supply.					
7	Negative attitude to use of the website					
8	High illiteracy rate among citizens.					
9	Financial resources for ICT.					
10	Website connectivity devices like MODEMS.					
11	Web presence where information can be posted.					
12	Training staff members to enable them use the website.					
13	Linking the website to websites of other local governments.					
14	Sensitizing citizens on the relevancy of the website					
15	Building trust in citizens by providing security to information.					
16	Putting in place other ICT facilities					

Chapter 3
Development of an Inclusive Participatory Democracy System

Aderonke A. Oni
Covenant University, Nigeria

Charles K. Ayo
Covenant University, Nigeria

Ambrose A. Azeta
Covenant University, Nigeria

ABSTRACT

Barriers such as unequal access, lack of digital skill, low income and disability constitute limiting factors for technology-mediated citizens and government interaction in developing countries. It is against this backdrop that this work explores the integration of Voice, Web and SMS technologies, in Nigeria's democratic process. The proposed system takes advantage of the ubiquitous nature of mobile devices to explore the plausibility of increasing the level of citizens' participation in democratic practices, particularly, those in rural areas with no Internet access and the physically challenged electorates. The server module for the e-democracy system was developed in PHP. Ozeki SMS server and Voxeo Voice server were used for SMS transaction code and VoiceXML code respectively. The prototype e-democracy system shows that developing nations can take advantage of their present level of technological development to give voice to the voiceless and improve their democratic system.

DOI: 10.4018/978-1-5225-2565-3.ch003

INTRODUCTION

The Global public management efforts since the 1980s have been revolutionizing the relationship between citizens and the State. The revolution focuses on governments being more responsive to the populace and requires new strategies to rebuild the relationship between citizens and governments and to promote more citizen participation in public administration (Kettl, 2005; Oni, et al., 2014). The use of Information and Communication Technology (ICT) in enhancing citizens' political participation has been identified as key to the revolution and a solution to the problems of representative democracy, particularly, the disconnection between representatives and citizens and the decline of political interest amongst the populace (Kang & Dugdale, 2010; Azeta, et al., 2015). E-participation initiatives have served as means for two-way communication between government and citizens. To the government, it serves as information provisioning tools and to the citizens as an avenue to voice their opinions, deliberate on policy issues, and give viable contributions to policy options. ICT has been persuasively argued to play a key developmental role in developing countries (Sahay & Avgerou, 2002; LeBlanc, et al., 2004). Leveraging on the capability of the internet and mobile technology, e-participation has the potential of creating new forms of engagement, deliberation, and collaboration in the political process to make democratic processes more inclusive and transparent (Coleman & Gotze, 2001; OECD, 2003b; Shirazi, et al., 2010). It has the potential to transform uncompetitive industries and dysfunctional public administration and to provide unprecedented opportunities for the information-intensive social services (Sahay & Avgerou, 2008).

Most countries have leveraged on ICT tools to foster new relationships between citizens and the State in order to alleviate the crisis of democratic legitimacy. The use of ICT is therefore, capable of bridging the disconnect between citizens and their representatives. Contrarily however, in developing countries such as Sub-Saharan Africa (SSA) countries, e-democracy implementation is still in its introductory stage, though most governments have embarked on e-government implementation strategies. In these countries, e-participation in democracy is mostly centered on private organizations' initiative. Considering the prevailing challenges of poor governance, lack of accountability and transparency, these governments mostly need to join their world counterpart in promoting citizens participation public decision making and explore the capability to put in place adequate infrastructure for e-democracy implementation.

After the start of the democratic era in Nigeria and the election of a democratic government dated as early as the 1990's till now, the elected government formed the National Empowerment and Development Strategy (NEEDS), the National e-Government Strategy (NeGST) and the National IT policy. Among others, the objectives of the Nigeria Information Technology (IT) policy include to improve accessibility to public administration for all citizens and to bring transparency to government processes. The strategies, among others, involve bringing the government to the doorsteps of people by creating virtual forum and facilities to strengthen accessibility to government information and facilitate interaction between the governed and the government leading to transparency, accountability and strengthened democracy (NITDA, 2001). Despite the establishment of strategies and national policies since 2000, Nigeria government is yet to fully explore the potentials offered by ICT to transform its political landscape. Given the proliferation of smart phone and mobile data access, Nigeria stands to benefit from using this technological development to advance its democratic system.

The exponential growth in the use of internet, mobile phone and especially mobile data in Nigeria within the last decade stand as platforms through which citizens can be engaged in public participation given available e-participatory channels. Nigeria was referred to as the fastest growing telecommunication country in Africa (Ayo et al., 2007). The number of internet users in Nigeria has continued to be on the increase. The number has grown from 200,000 Internet users in the year 2000 to 43,982,200 i.e., 28.3% of the country's population and 70,300,000 (39.7% of the country's population) in 2014 (www.internetworldstats.com/africa). The increasing numbers of internet users in Nigeria from 0.1% in 2014 to 39.7% of its population in June 2014, revealed that the use of internet in the country is growing at a sporadic rate and still has the potential to grow higher. The 2012 World Internet statistics also revealed that internet usage in Nigeria increases with increase in population. Likewise, mobile technology has gained wider acceptance all over the country. Mobile phone usage statistic of March, 2014 revealed that there are 138,960,320 (78.44% of Nigeria population) mobile phones in use in Nigeria (www.ncc.gov.ng). Fifty percent of internet users are mobile web users who never or infrequently use the desktop web (www.mobithinking.com). With the wide acceptance of mobile technology and the available internet connections, Nigeria has the potential of employing ICT intervention in addressing her democratic problems.

It is on this basis that this study aims to propose an inclusive e-democracy system for sustainable citizens' participation in public decision making in Nigeria. The challenge to be considered is how to develop a suitable multimodal e-democracy system incorporating some e-participation tools which can help cater for the marginalized group especially those without internet access and the visually impaired. Leveraging on the capability of internet and mobile technology and lessons from e-democracy implementation around the world, this work demonstrates the plausibility of an inclusive citizen e-participation in democracy in resource-poor settings like Nigeria. The importance of ICT in transmitting e-democracy through mobile technology is numerous (Ayo et al., 2012). It facilitates dissemination of information about politics and discussion of political issues. Wider democratic participation by individuals and groups and greater transparency and accountability in democratic institutions and processes are indeed its benefit to citizens and the society.

BACKGROUND

Beside the sovereign feature of the people governed, democracy as a political system has other identifiable features which distinguish it from other systems of governance and also demands that citizens political participation should go beyond voting to active participation in public making (Coleman & Gotze, 2001; OECD, 2003a). These include among others:

- **Citizen Participation:** In a democratic context, the more the citizens participate the more legitimate the outcome of democracy (Chen, et al., 2006). Participation is not only citizens right, it is their duty. The act of participation may take several forms including voting in elections, becoming informed, debating issues, attending community or civic meetings, being members of private voluntary organizations, paying taxes, protesting against undesired decision and standing for election (Schumpeter, 1976; Coleman & Gotze, 2001).
- **Equality:** Another basis of democracy is the principle of equality. The theory and practice of democracy is built on the concept of human equality. Equality means giving equal opportunity to all, rating all individuals equally without discriminating against race, religion, ethnic group, gender or sex (democracy for All). Al Amer (2009) identified five kinds of equality as Civil equality, Political equality, social equality, natural equality, economic equality.
- **Liberty:** This is another key democratic principle. Liberty is the state of freedom. According to Al Amer (2009), two major categories pertaining to liberty are political freedom, and freedom of expression and organization. Political freedom guarantees citizens the right to participate freely and without fear and, include voting, standing and campaigning for office (Jahanbakhsh, 2001). Political freedom also includes the freedom of the citizens to express their claims publicly, to criticize decision-makers and propose alternative policies (Al Amer, 2009). Freedom of expression and organization include citizens' freedom to form associations and assemblies to actualize their political goals without any fear of punishment (Jahanbakhsh, 2001).
- **Political Tolerance:** Democratic system presumes the societies to be politically tolerant. This means that while the majority rules, the rights of the minority is protected. People who are not in power must be allowed to speak out.

- **Accountability:** Political officials must make decisions and perform their duties to satisfy the will and wishes of the people, not of themselves. Elected and appointed officials have to be accountable to the people and as well take responsibility for their actions.
- **Transparency:** Transparency in government is paramount in democracy. For a government to be accountable, it must keep citizens abreast of all its decisions and policies.

In time past, elected representative was a sufficient means for citizens to participate in government (Smith, 2003). This has changed for reasons such as diversity of citizens' cultural heritages, values, needs, and apathy, lack of openness and accountability, and mistrust prevailing in democratic societies (Smith, 2003). Citizens now want to be broadly involved in governance and policy decisions. But this is not achievable through the traditional mass media such as television, radio, or newspapers. Traditional public participation mechanisms such as citizen advisory committee, public hearing, planning cells, petition, referendums, focus groups and citizens' juries are usually dominated by special interest groups - organized bodies/association controlled the smaller groups. Feed-back obtained may not be representative of the community. In addition, the inarticulate and perhaps disadvantaged groups are excluded. Only minorities which belongs to the highly educated groups, representative of parties, members of higher job-related positions, public services participate. The need for more avenues for citizens participation in governance necessitates e-participation.

Combining the words "Electronic" and "Democracy", according to Caldow (2004), will mean the use of ICT tools to facilitate, improve and extend democratic activities. This has the potential of creating new forms of engagement, deliberation, and collaboration in the political process to make democratic processes more inclusive and transparent (Coleman & Gotze, 2001; OECD, 2003b; Shirazi, et al., 2010). E-democracy is, therefore, anything that governments do to facilitate greater participation in government and enhance effective governance using digital or electronic means (Colman & Norris, 2005). According to Cliff (2003), e-democracy is considered as the use of information and communication technologies and strategies by democratic actors (governments, elected officials, the media, political organizations, citizen) within political and governance processes of local communities, nations and on the international stage. The principal ICT mechanism used in e-democracy is the internet accessed through various channels, including PCs, mobile phones, and interactive digital TV both in the homes and in pubic locations.

According to Caldow (2004), e-democracy has two sides: the tactical side and the strategic side. The tactical side deals with the use of information technology to advance communication and promote access to information while the strategic side considers how government can use digital media to actively engage citizens and advance its public policies to the world community. This includes the use of e-democracy initiatives to engage citizens in democratic processes such as policy deliberation, information enquiry, petition against corrupt practice, and expression of opinions through forums (Coleman & Gotze, 2001; Funikul & Chutimaskul, 2009). The Organization for Economic Co-operation and Development (OECD) stated that citizens' democratic political participation must involve the means to be informed, the mechanisms to take part in the decision-making and the ability to contribute and influence the policy agenda (OECD report, 2001). The organization, therefore, proposed a three stage levels of citizens' engagement in e-democracy to include:

- **Information Provisioning:** This is a one-way relationship in which government provides information for citizens. Examples include: access to public record, official gazettes, and government Web sites.
- **Consultation:** This is a two-way relationship where citizens have opportunity to give feedback on issues. Government defines issues on which citizens' views and opinions are required and also involved to manage the process. Examples include comment on draft legislation and opinion survey.
- **Active Participation:** It is a partnership based relationship where citizens are actively engaged in defining the process and content of policy-making. It operates on the bases of equality among citizens in proposing policy options, amendment and deliberation, although final decision rests in the arms of government. Popular examples include citizen juries, consensus conference, etc.

Notable efforts have been made by governments in some advanced countries to address the issue of inadequate public participation in democratic processes. Most of these efforts centered on the use of internet. Advanced democracies such as Australia, United Kingdom, Scotland, Brazil, United States and Korea are now using technology to gather input from citizens. Public participation techniques currently being used to harvest citizens' input include electronic discussion groups for the deliberation and development of policy options, online referenda, online opinion polls, and e-petitions (to government or parliament) (UN e-government Survey, 2008; Caldow, 2004; OECD, 2003a). Citizen's engagement recognizes citizens as stakeholders and seeks to involve them directly and at a depth not achieved through traditional methods of public consultation, and group representation provided by various interest groups and associations (Smith, 2003).

There are several benefits that government can derive in making use of electronic techniques for public participation. Smith (2003) and Clift (2004) enumerated some of the benefits to include:

- **Enhance Public Knowledge, Understanding, and Awareness:** Sharing information about issues of public concerns gives opportunities for stakeholders to hear and understand each other views.
- **Meet a Growing Demand for Public Participation:** The clamor of every society for democratic system of governance is to participate in decision making process and as well, to have openness of decision making processes.
- **Solidifies Legitimacy:** Public participation would help to reduce public mistrust of the government.
- **Enhance Effectiveness:** Participation produces higher quality decisions. Considering all relevant information, views, needs, and interests in complex decision help policy maker to get it right and secure public commitment.
- **Public Participation:** Also helps to resolve conflict, negotiate tradeoffs, establish priorities and seek consensus.

Despite these advantages and the advocacy by societies, scholars and governments for informed and engaged citizenry through the use of ICT, the problem of digital divide resulting from citizens' unequal access, lack of proper infrastructure and low adoption of technology have been argued to limit the perceived benefits of e-participation in developing countries (Rheingold, 2000:377; OECD, 2001:15; OECD, 2003b:60; Rodousakis & Mendes dos Santos, 2008). Barriers in technology-mediated citizens and government interaction include, among others, awareness, access, skills and attitude. Each of these barriers may lead to e-exclusion, though socially disadvantaged persons often suffer from multiple deprivations (Rodousakis & Mendes dos Santos, 2008).

Accessibility being one of the critical success factors of e-democracy implementation, it thus connotes that e-democracy implementation, especially in the resource-poor countries, should be driven using a strategic approach guided by quality designs to enhance sustainability and make tools for online engagement accessible to all segments of the society irrespective of barriers such as age, disability, low level of education, lack of digital skill, and low income. It should also give the socially underprivileged access to engage in e-democracy via complementary channels (Rodousakis & Mendes dos Santos, 2008, Ayo et al., 2012). It is against this backdrop that this work aims at exploring the use of internet, mobile and voice technologies, particularly the mobile devices, in Nigeria's democratic process. The rationale is to make tools for e-democracy available on a platform that is most accessible to the populace. It has been observed that the wide spread of ICTs particularly, mobile platform can be harnessed to enhance better participation in the polity. A multi-modal e-democracy application is being put in place to take advantage of the numerous mobile phone users in developing countries.

PROTOTYPE IMPLEMENTATION OF AN INCLUSIVE E-DEMOCRACY SYSTEM

This section gives a brief description of the prototype implementation of the e-democracy architecture shown in Figure 1. It is a 3-tier architecture comprising of client layer, middleware and data tier. The client layer consists of users' available medium of connectivity such as phone, PC, tablets, etc. The client-server interaction is based on standard internet protocols Hypertext Transfer Protocol (HTTP) and Transmission Control Protocol/Internet Protocol (TCP/IP). The middle tier comprises of the speech server, the web server, the SMS server and the application server.

Since harnessing citizens input in public decision making irrespective of their technological access is key to inclusive e-participation, the application layer consists of e-participation tools including e-petition, e-forum, online opinion poll, mobile SMS, e-consultation and information. These tools are implemented for two categories of user, the voice application users and the web application users except the opinion poll which includes SMS platforms. Data layer consists of data services and database managements functions. The data layer is a necessity to the Voice Mobile Democracy system because it contains its vital information and contents of the system i.e., pre-recordings of news and polling questions. MySQL database is used for data layer.

Figure 1. Proposed e-democracy software architecture

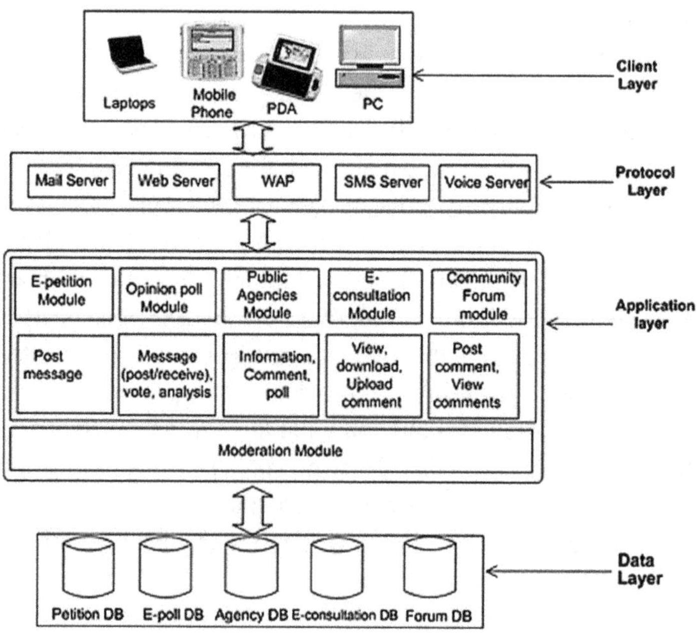

Description of System Components

The proposed e-democracy system contains the following components

- **E-Consultation:** This tool provides a platform for consultative interaction between government and citizens. It is developed to seek public opinion on proposed plans, policy draft or implementation plan. It provides citizens opportunity to give their opinion more objectively than some of other tools. The system provides list of available new initiatives, policy draft, and government policies with the background information and submission closing dates. Since each consultation issue is given ample closing date and adequate information on government proposal, citizens can contribute significantly to improving or abolishing such proposal. Each observation suggestion and comments to be provided by citizens has maximum word limit to avoid irrelevance and ensure that the opinion is constructive and concise.
- **Online Opinion Polls:** This allows the citizens the opportunity of adding a voice to any issue of importance. Opinion poll is used to conduct polls on a particular issue, in order to see which of the available options the general public agrees with. The system administrator posts the subject matter and available options and then user vote by choosing among available options. It offers decision makers quick access to knowing the choice of the populace and guides their decision making.
- **Mobile SMS:** This is the SMS based version of the online opinion poll. It handles outgoing and incoming short message service. The administrator can send SMS to mobile phone number of different categories of stakeholders on the system through the SMS server. Response format is indicated in the outgoing SMS so that the system can automatically analyze the result.
- **E-Forum:** The forum enables the participation of citizens in public discussions. This module will help launch new fora effectively by attracting the participation of active citizens, elected officials and policy makers, interest groups, and journalists to deliberate and provide recommendation to solving bordering public issues.
- **E-Petition:** This platform is used to capture complaints of citizens and seeks redress on issues of concern. On this platform, citizens can send requests to the authority on a variety of local and national matters such as food, security, housing, transportation, importation, etc. A citizen fills e-petition form and after it has been moderated to ensure that it conforms to the e-petitions rules and guideline, it is made public and other citizens can sign it. Every petition with signature that has reached the minimum specified threshold with the allotted time must be given consideration in the parliament. The outcome of

every decision on such petition is made available online. The importance of this is for the public influence on policy, to improve public understandings and awareness of the parliament's work and strengthen democracy.

- **Information Provision:** This module provides information on ongoing political issues and government activities to the public. It is also meant to facilitate citizens access to public record. The essence is to provide access to information and data that cannot be gotten through traditional mass media such as television and radio.
- **System Administrator:** This module is only available to the system administrators/moderators whose responsibilities are to ensure effective coordination and compliance. It involves system maintenance, general management of the system and moderation of users activities. Moderation is very important in e-participation to ensure that user operates within the rules guiding the operation of each e-participation tools.

System Modeling

A commonly acceptable tool for modeling the functionality of a system is the Unified Modeling Language (UML) tools. UML tools used for modeling the functionalities of the e-democracy system include: use case, activity diagram and sequence diagram. This section describes the functionalities of the system using the tools specified:

Activity Diagram

An activity diagram is an important UML diagram that describes the dynamics of a system. It depicts the flow of events in a system while modeling the operations of the system. Its purpose is to capture dynamic system behavior and to construct the next executable event after completion of an action. Figure 2 is the activity diagram for a citizen viewing and responding to an e-consultation. Every visitor to the system can view available consultations and also download additional information provided in the attachment. A user has to be a registered user and login to participate in the e-consultation.

Figure 3 is the Use case diagram of the e-petition module; users can view existing e-petition, create a new e-petition or sign an open e-petition by adding their name and address.

Voice User Interface (VUI) Design

VUI design is a type of human computer interaction design which requires the ability to represent or design voice input data and grammar such that it can be easy

Figure 2. Activity diagram for responding to an e-consultation

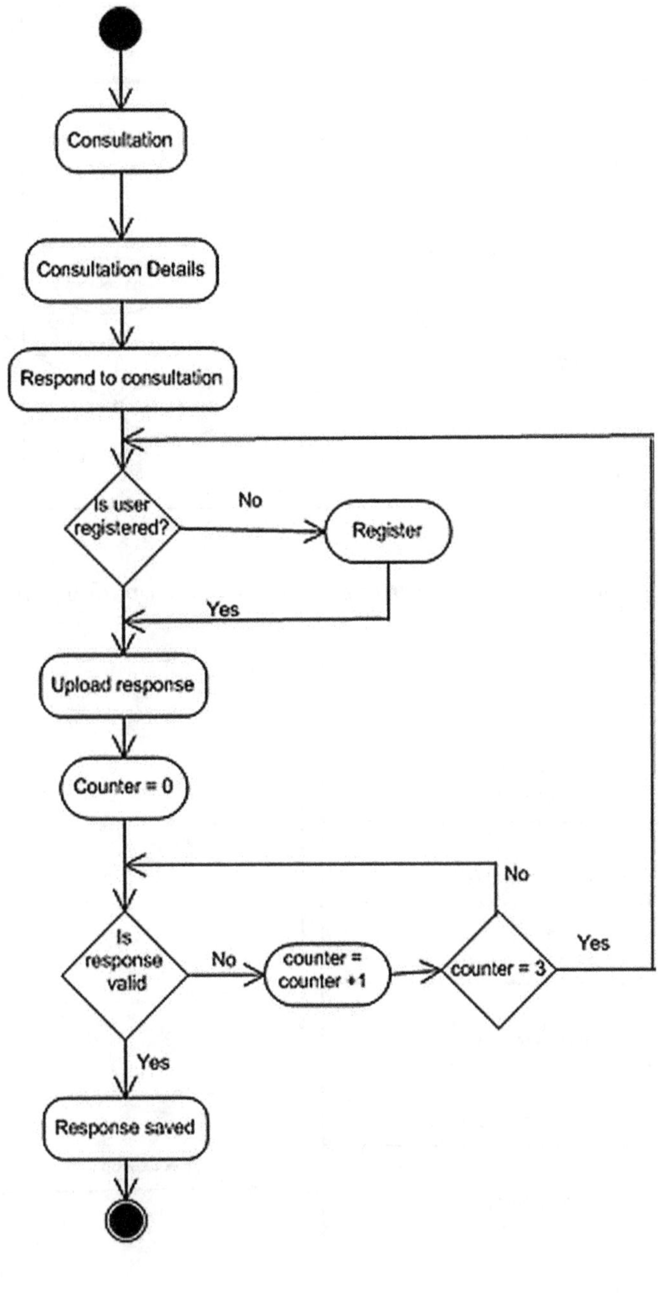

to navigate through system and flexible enough for usability purposes. In the course of this design, exception handling for different activities was most considered, as the voice server throws exceptions when users finish supplying inputs to system and

Figure 3. Use case for signing and creating new e-petition

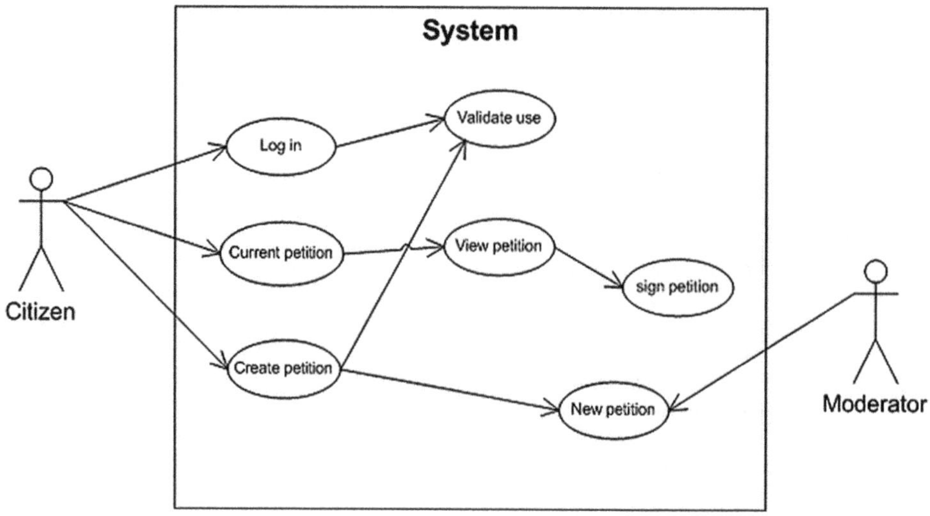

Figure 4. Sequence diagram of user interaction with the opinion poll voice module

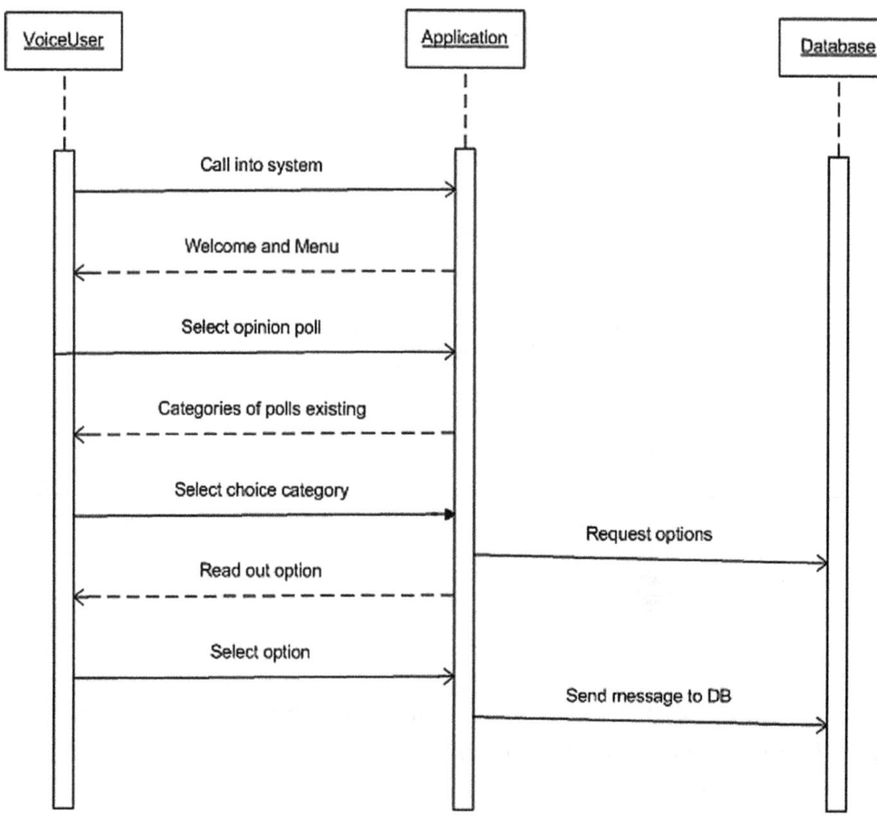

most especially, when there are mismatches between users input and the required input. System gives users opportunity to try again and after several attempts, if user fails to give appropriate input, program terminates.Figure4, is a sequence of call flow voice user and the e-democracy system

Mobile SMS Interface Design

The SMS module is used for sending SMS alert message to group of citizens on opinion poll and other pertinent news. Citizens can use this module to participate in public opinion poll by sending their response according to predefined format.

IMPLEMENTATION

The prototype system consists of several modules implemented in three different platforms, Web user interface, voice interface and mobile SMS. It was developed on open source platform using PHP 5.2, MySql server, Apache 2.2.1, and MyODBC 3.51 driver to link the SMS server and the MySql server. Ozeki SMS server was used to deploy the SMS transaction code for sending and receiving of text messages through the application. Voxeo Voice server was used to deploy the VoiceXML code.

The voice users communicate with the voice server by dialing a number on mobile or land phone. User's voice input is translated to text by Automated Speech Recognition (ASR) system and the voice server communicates back to the user by using Text To Speech (TTS).

E-Consultation Module

Figure 5 shows the screen shots of the page showing all available consultation in the e-consultation module. The module is intended to consult the public for opinion on proposed plans or any action that is about to be implemented. Adequate information is provided on the subject area and people can give their comments or observations on the issues using the form provided. Figure 5 displayed a list of available consultations with the name of the departments or institutions seeking public contribution, and the date the consultation will close for public contributions. Clicking any of the consultation topics displays the background information to the consultation which includes a download as it may be required. The user can proceed to fill the e-consultation form to participate in the consultation. Each fields of contribution has word limit in order to avoid irrelevances.

The E-Petition Module

This module is to be used by citizens to send a written appeal to parliament on public issues to which redress is required. Every e-petition submitted goes through moderation to ensure its conformance with terms and conditions before it is published online for other citizens to sign thee-petition by adding their names and addresses online.

The first page on the e-petition module provides a brief introduction about e-petition, who is eligible to petition government/parliament, how the e-petition works and a link to the terms and conditions attached to e-petition. Every e-petition sent must adhere strictly to the terms and conditions of e-petition otherwise, it will be rejected. When a user sends e-petition, the moderator checks the e-petition to confirm that it does not violate any rule of the system such as “no abuse of person”. The petitioner is only expected to write what the issue of concern is and the action desired to be taken in a limited number of words. After agreeing to the terms and conditions of e-petition, users can then see available e-petitions as shown in Figure 6. The figure shows a list of open e-petitions containing the subject of the e-petition, senders’ names, time sent, and the date the e-petition will close for signing.

E-Forum Module

The module allows interaction among citizens in form of public discussion. A citizen creates topic and other citizens comment on it. The administrator is given the privilege to moderate the content of a topic completely if it violates participation principles such as no abuse of persons.

Figure 5. Screen shot showing list of open consultation

Figure 6. Screen shot of open e-petition

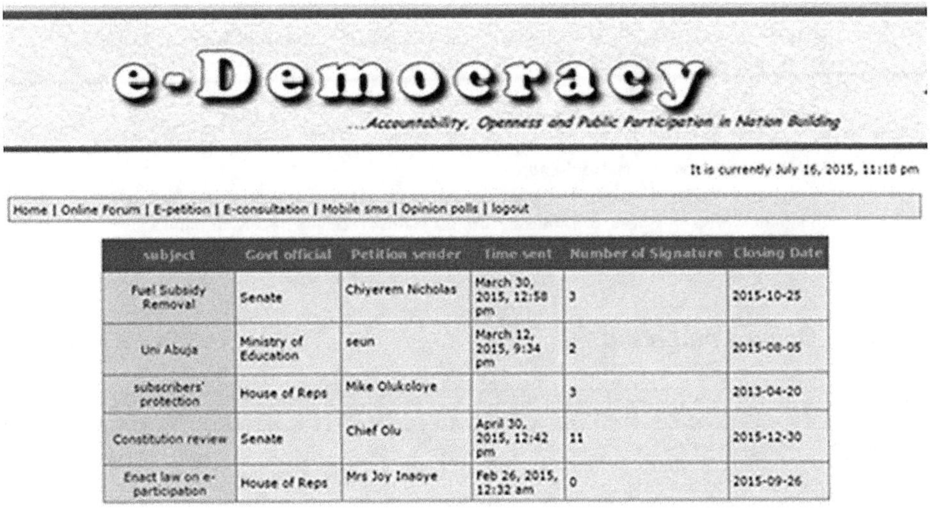

subject	Govt official	Petition sender	Time sent	Number of Signature	Closing Date
Fuel Subsidy Removal	Senate	Chiyerem Nicholas	March 30, 2015, 12:58 pm	3	2015-10-25
Uni Abuja	Ministry of Education	seun	March 12, 2015, 9:34 pm	2	2015-08-05
subscribers' protection	House of Reps	Mike Olukoloye		3	2013-04-20
Constitution review	Senate	Chief Olu	April 30, 2015, 12:42 pm	11	2015-12-30
Enact law on e-participation	House of Reps	Mrs Joy Inaoye	Feb 26, 2015, 12:32 am	0	2015-09-26

Opinion Poll

This module is used to conduct polls. It has web and mobile SMS platform. Figure 7 shows the SQL transaction code for sending and receiving via the messages SMS platform. Citizens with web access upon login, can participate in the opinion poll and view the report as shown in Figure 8. In the same vein, the administrator uses the web platform to send messages by category to citizens and government officials. Different categories of public office holders and citizens based on their location exist in the database. To send SMS to a particular group, the system administrator simply select the group and the application picks the corresponding phone numbers from the database and runs SQL query to send the intended message to the phone numbers in the selected category. Incoming SMS messages go directly into the system database and the result is computed with the web based opinion poll.

Voice Application

Voxeo Prophecy version 14 was used to develop the voice application. The application, so far, contains three modules: online information, opinion poll, and e-petition. These three modules work just like the web application. Clicking dial button on the SIP Softphone keypad (see Figure 9) after typing the application's name launches the application's welcome message, it then proceeds to authenticate user. Once identification is completed, the system presents the main menu.

Figure 7. SQL transaction code for sending and receiving SMS

```
Receiving incoming messages: insert into ozekimessagein (operator, sender, receiver,msg,
senttime,receivedtime,msgtype,reference) values ('$operator', '$sender','$receiver','$msg',
'$senttime','$receivedtime','$msgtype','$reference')

Sending Out Messages from the Database
Download messages: select id,receiver,msg,operator,msgtype,sender,reference from
ozekimessageout where status='send'
Update downloaded message: update ozekimessageout set status='sent' where id='$id'
update delivered messages: ozekimessageout set status='received',receivedtime=
'$recvtime' where id='$id'
```

Figure 8. Opinion poll result

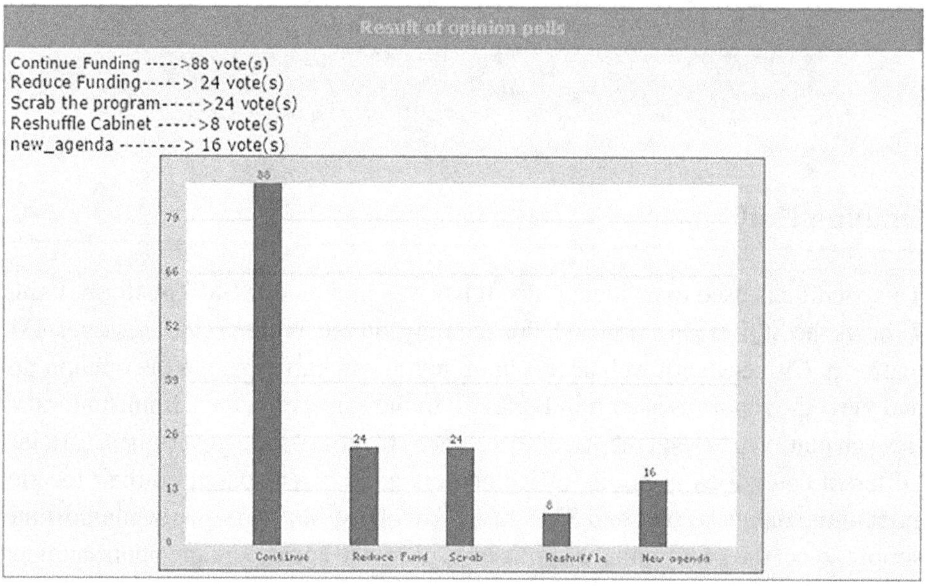

EVALUATION OF THE PROTOTYPE SYSTEM

This section presents the evaluation of the prototype e-democracy system. The purpose of this evaluation is to assess the design and usability of the system. The evaluation was conducted based on the socio-technical criteria for e-participation proposed by Macintosh and Whyte (2008).

A post-experience questionnaire was developed based on the socio-technical criteria including trust and security, relevance, usefulness, usability, content clarity, error recovery and flexibility for e-participation. The questionnaire is divided into two parts. The first part required the participants to rate their level of understanding

Figure 9. Voxeo Prophecy SIP phone

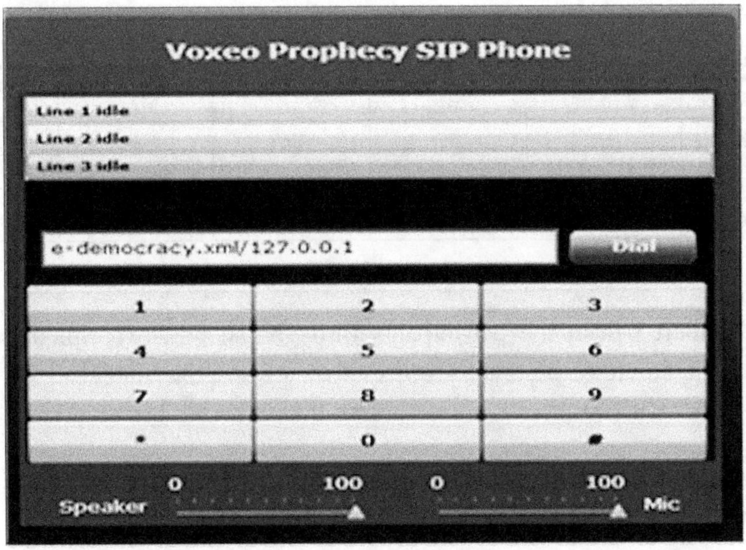

of and experience with the use of the e-democracy system. The second part of the post-experience questionnaire contains two questions on each of the seven items of the socio-technical criteria for e-participation. The questions adapted to fit the scenario of this study were developed for each of these seven items based on the description of Macintosh and Whyte (2004). The participants were required to rate each item on the questionnaire on a scale of 1-5 (1- Strongly Disagree, 2 – Disagree, 3 – Indifferent, 4- Agree, and 5 – Strongly Agree).

A total of twenty (20) participants consisting of seven (7) undergraduate students and thirteen (13) postgraduate students were selected to carry out the evaluation. The participant were selected from the Department of Computer and Information Sciences, College of Science and Technology and Department of Political Science and International Relations, College of Human Development, Covenant University, Nigeria. Selection of this group of people is based on their proficiency in Internet application usage and ability to contribute meaningfully to political matters. All the participants are existing users of social media such as facebook, twitter and Linkedln which makes us believe that they are capable to evaluate this system. The selected participants were given an overview of the e-democracy system. The functions of various modules in the system was explained to them and they were allowed to personally experience the system to know the extent to which an individual could easily work-through the system. After receiving feedback on their first experience with the system, they were given the post-experience questionnaire.

Table 1. Participants understanding and experience with the use of e-democracy system

Understanding	Frequency	Percent	Experience	Frequency	Percent
Very High	5	25	Very High	2	10
High	11	55	High	9	45
Medium	4	20	Medium	9	45
Total	20	100	Total	20	100

Data gathered from the participants through the post-experiment of socio-technical evaluation of the system was analyzed using the Statistical Package for Social Science (SPSS).

Assessment of Understanding and Experience with the E-Democracy System

The participants were asked to indicate their level of understanding and experience with the use of the e-democracy system as Very High, High, Medium, Low or Poor. Table 1 presents the participants responses of their understanding and experience with the system.

Rating of the System by Participants

Table 2 shows the descriptive statistical analysis of participants evaluation of the e-democracy system using the socio-technical critieria.

- **Trust and Security:** This involves the security of users' information and the quality of information provided on the system. Specifically, it involves accuracy, reliability and completeness of the information in the system. The mean rating for this item is 4.05. The frequency graph is shown in Figure 10. It shows that 50% of the participant agreed that the e-democracy system is secured and can be trusted.
- **Relevance:** Relevance, according to Macintosh and Whyte (2008), involves users' satisfaction that the system provided meets a purpose relevant to their political needs and that the contents and surrounding processes are relevant to the purpose. The mean rating for this item is 3.89. The frequency graph in Figure 10 indicates that 60% and 20% of the participants agreed and strongly agreed, respectively, that the e-democracy system is relevant to their political needs.

Table 2. Descriptive statistics of post-experience data

S/N	Criteria	Mean	Std. Deviation	Variance
1.	Trust and Security	4.05	0.536	0.287
2.	Relevance	3.97	0.697	0.523
3.	Appeal	4.13	0.887	0.786
4.	Navigation and Organization	3.73	0.835	0.697
5.	Content Clarity	4.30	0.470	0.221
6.	Efficiency and flexibility	4.15	0.564	0.318
7.	Error Recovery	3.48	1.057	1.118

- **Appeal:** The mean rating for this item is 4.13. The item examines the usefulness of the system. It concerns with take-up in line with expectation (Macintosh and Whyte, 2008), users desirability of the system and their intention to use it again. The frequency graph of the items shows that 20% of the participants agreed that the system is appealing while 45% indicted that the system is strongly appealing.
- **Content Clarity:** Content clarity is also related to usefulness of the system. It measures users understanding of the content of the system in relation to the task to be performed. The mean rating of the contents clarity in this evaluation is 4.30. The frequency graph in Figure 10 shows that 70% of the participants agreed that the contents are clear and the other 30% strongly agreed that the contents are clear.
- **Navigation and Organization:** The mean rating for navigation and organization is 3.73. Navigation and organization, efficiency and flexibility and error recovery are usability assessment measures. Navigation and organization assess the sufficiency and consistency of information available to users relative to their current position within the site organization, the path taken and the options available. The frequency graph of users' evaluation of navigation and organization of the system (Figure 10) shows that 35% and 25% of the participants agreed and strongly agreed, respectively, that the system contains sufficient and consistent information about current position and also contains active links.
- **Efficiency and Flexibility:** The mean rating of efficiency and flexibility is 4.15. Efficiency and flexibility assess the possibility of users being able to perform tasks in an acceptable time. It also examines the availability of appropriate short-cut for performing repetitive tasks. The frequency graph of users assessment of the efficiency and flexibility of the system (Figure 10) shows that 70% and 20% of the participants agreed and strongly agreed,

Figure 10. Frequency distribution of user evaluation

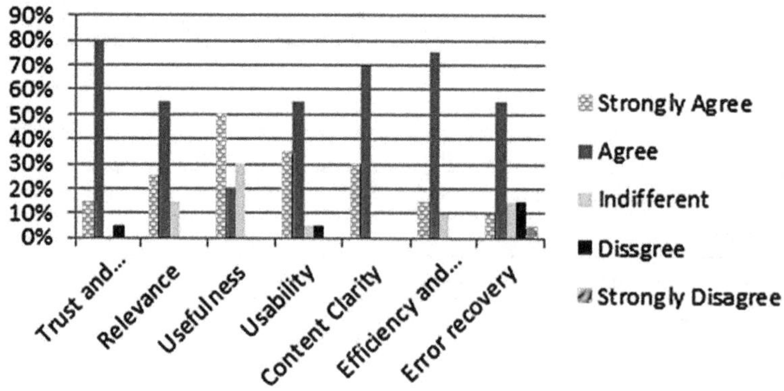

respectively, that the system saves time and requires few steps to accomplish any task.

- **Error Recovery:** Error recovery deals with the ability of the system to allow users to "undo" previous action and be guided effectively on the correct procedure so as to be able to complete the task without distraction. The mean rating for error recovery is 3.45. According to the frequency graph in Figure 10, 35% of the participants agreed that the system allows error recovery and only 15% strongly agreed that it allows error recovery.

SUMMARY AND RECOMMENDATION FOR FUTURE WORK

This chapter has presented the implementation and evaluation of the prototype e-democracy system developed to realize inclusive political participation of citizens in developing countries and Nigeria in particular. The prototype multimodal e-participation system is premised on internet, mobile, and voice-based technology. The chapter, firstly, reviewed the characteristics of democratic system of government that necessitate the need for inclusive political participation. It went further to explain the concept of e-democracy, various stages of e-participation in polity and the attendant benefits of e-participation. The prototype e-democracy system was developed to demonstrate, among others, the plausibility of engaging citizens in democratic process at the highest level of engagement (active participation) using some of the e-participation tools. The evaluation of the e-democracy application was undertaken in order to assess the quality of its system design and usability. The evaluation was conducted based on the socio-technical criteria for e-participation proposed by Macintosh and Whyte (2008). The criteria addressed various design aspects including trust and security, relevance, usefulness usability, content clarity, error recovery and flexibility. The result of the evaluation clearly revealed that the users found the system flexible to interact with. The contents and organization of the system are easy to comprehend. The system is found to be appealing and relevant to the political needs of the users.

This study provides opportunities for further research in e-democracy. The prototype system can be enhanced by semantic web technology and data mining technology. Semantic web technology will offer opportunity for processing information via the internet with the help of machines which are capable of handling semantics of information from heterogeneous sources. Incorporating this into the present e-democracy system will enhance search and provision of information on the system. It will also provide argument visualization mechanism. Data mining technologies will provide opportunity for knowledge discovery. It will be useful in this context to evaluate and extract valuable knowledge from large discussion data. The voice-based part of the e-democracy system should be further evaluated for users' satisfaction and memorability of the system. This evaluation could also be complemented with some feature of PARAdigm for Dialogue System Evaluation (PARADISE) (Walker et al., 1997) that could be adapted to this system.

CONCLUSION

While some developed countries have been exploring ways of engaging their citizens in online consultation with government officers and thus, allowing them to give their opinions on policy options, countries in Sub-Sahara Africa have not adequately explored the intervention of ICT in integrating citizens' input into policy-making process (Coleman & Gotze, 2001; OECD, 2001; Caldow, 2004; Kim, 2008). This research is therefore timely and significant in the following areas:

This study has the potential to boost the involvement of citizens in democratic processes, thereby making citizens more responsive to national development through policy deliberation and active participation in public decision making process. This study has the potential of improving the credibility of a nations' legislative process by making policy makers open and more responsive to public opinions. The study is also capable of ameliorating most of the political impasse in the developing nations of the world, particularly in the African State. The prototype e-democracy system developed will be useful for e-participatory democracy without prejudice to disability thereby giving opportunity for all and sundry to be involved in democratic governance. Also, government agencies can raise power balance when connecting directly and efficiently with citizens instead of going through elected officials or the mass media (Clift, 2004). The proposed system has the potential to increase the level of participation of citizens in government.

Potential exists for e-democracy to increase political participation in developing democracies. The benefits, enablers, and barriers associated with e-democracy should be given due consideration when exploring its implementation and impact on the democratic process. Active engagement of citizen in democratic process through ICT can reduce the disengagement between the representative and the electorate. The role of democratic citizens should not stop at voting but stretch to deliberating to maximize the benefit of representative democracy.

The proposed system explores the integration of voice, web and SMS technologies to reduce access barriers to e-participation in democruatic governance. The system takes advantage of the ubiquitous nature of mobile devices to increase the level of citizens' participation in democratic practices particularly the physically challenged electorates. Similarly, the integration of voice, web and SMS technologies makes it suitable for citizens in rural areas and the visually impaired to participate in e-democracy regardless of their technological access or level of disability. This multimodal e-participation system approach will enhance e-democracy implementation and facilitate a wider access to citizens' participation in governance most especially in developing countries. This study concludes with a brief demonstration of the plausibility of an inclusive e-democracy implementation for resource poor nations like Nigeria. This proof concept shows that developing nations can take advantage of their present level of technological development to give voice to the voiceless.f

REFERENCES

Al Amer, M. A. (2009). *Electronic Democracy Strategy for Bahrain* (Unpublished doctoral dissertation). De Montfort University, Leicester, UK.

Ayo, C. K., Azeta, A. A., & Oni, A. A. (2012). E-Democracy: An Enabler for Improved Participatory Democracy. In K. J. Bwalya & S. F. Zulu (Eds.), *Handbook for e-Government in Emerging Economies - Adoption, Continuance Usage, e-Participation and legal frameworks*. IGI Global. doi:10.4018/978-1-4666-0324-0.ch018

Ayo, C. K., Ekong, U. O., Fatudimu, I. T., & Adebiyi, A. A. (2007). M-Commerce Implementation in Nigeria: Trends and Issues. *Journal of Internet Banking and Commerce, 12*(2).

Caldow, J. (2004). *E-democracy: Putting Down Global Roots. Institute of Electronic Government*. IBM.

Azeta, A., Azeta, V., Olaniyan, O., Azeta, A., & Ayeni, A. G. (2015). Implementing an E-Democracy System in Nigeria. *Journal of Resources Development and Management, 4*, 1–11.

Chen, D., Huang, T., & Hsiao, N. (2006). Reinventing Government Through On-Line Citizen Involvement in the Developing World: A Case Study Of Taipei City Mayors E-Mail Box In Taiwan. *Public Administration and Development, 26*(5), 409–423. doi:10.1002/pad.415

Clift, S. (2003). *E-Democracy, E-Governance and Public Net-Work*. Retrieved Jan., 2012 www.publicus.net/articles/edempublicnetwork.html

Clift, S. L. (2004). *E-government and Democracy: Representation and Citizen Engagement in the Information Age*. Retrieved Nov., 2010from http://www.publicus.net

Coleman, S., & Gotze, J. (2001). *Bowling Together: Online Public Engagement in Policy Deliberation.* Hansard Society. Retrieved 8th Oct., 2010 from www.hansardsociety.org.uk

Coleman, S. & Norris, D. F. (2005, January). *A new agenda for e-democracy*. Oxford Internet Institute Forum Discussion University of Oxford, UK.

Funilkul, S., & Chutimaskul, W. (2009). The Framework for Sustainable eDemocracy Development, *Transforming People. Process and Policy, 3*(1), 16–31.

Internet World Stats. (n.d.). *Usage and Population Statistics*. Retrieved from http://www.internetworldstats.com/stat1.htm /Africa

Jahanbakhsh, F. (2001). *Islam, Democracy and Religious Modernism in Iran,1953-2000: From Bazargan to Soroush*. Leiden, The Netherlands: Brill.

Kang, H., & Dugdale, A. (2010). A Comparative Study of e-Civil Participation in Australia and South Korea: The Case Study of GetUp! in Australia and the 2008 Candlelight Protest in South Korea, *Proceedings of 6th International Conference on e-government: Academic Conference International.*

Kettl, D. F. (2005). *The Global Public Management Revolution* (2nd ed.). Washington, DC: Brookings Institution Press.

LeBlanc, P., Mbarika, V., McCoy, S., & Meso, P. (2004). TeleDemocracy in developing countries: a focus on Sub-Saharan Africa.*Proceedings of the 2004 Information Resource Management Association (IRMA 2004).*

Macintosh, A., & Whyte, A. (2008). Towards an evaluation framework for eParticipation. *Transforming Government: People. Process & Policy*, *2*(1), 16–30.

Nigerian Communications Commission. (2015). *Subscriber Statistics*. Accessed July 7, 2015 from http://www.ncc.gov.ng

Nigerian National Information Technology Development Agency (NITDA). (2001). *National Policy for Information Technology*. Retrieved from http://portal.unesco.org/en/files/3107/1023717285 nigeriaitpolicy.pdf/nigeriaitpolicy.pdf

OECD. (2001). *Citizens as Partners Information, Consultation and Public participation in Policy-Making*. Retrieved Dec, 2010 fromwww.ezd.si/fileadmin/doc/4.../Citizens_as_partners_ hanbook_oecd.pdf

OECD. (2003a). *Engaging Citizens Online for Better Policy-making*. Retrieved Dec., 2010 from www.oecd.org/dataoecd/62/23/2501856.pdf

OECD. (2003b). *Promise and Problems of E-Democracy: Challenges of Online Citizen Engagement.* Retrieved Dec., 2010 from www.oecd.org/dataoecd/9/11/35176328.pdf

Oni, A. A., Ayo, C. K., Mbarika, V. M., Gberevbie, D. E., & Folarin, S. F. (2014). E-democracy Implementation: The Imperative of Agenda Setting.*Proceedings of 14th European Conference on E-Government.*

Rheingold, H. (2000). *The Virtual Community*. The MIT Press.

Rodousakis, N., & Mendes dos Santos, A. (2008). The development of inclusive e-government in Austria and Portugal: A Comparison of Two Success Stories. *The European Journal of Social Science Research*, *21*(4).

Sahay, S., & Avgerou, C. (2002). Introducing the Special Issue on Information and Communication Technologies in Developing Countries. *The Information Society*, *18*(2), 73–76.

Schumpeter, J. (1976). *Capitalism, Socialism and Democracy* (5th ed.). London: George Allen &Unwin Ltd.

Shirazi, F., Ngwenyama, O., & Morawczynski, O. (2010). ICT expansion and the digital divide in democratic freedoms: An analysis of the Impact of ICT Expansion, Education and ICT Filtering on Democracy. *Telematics and Informatics*, *27*(1), 21–31. doi:10.1016/j.tele.2009.05.001

Smith, B. L. (2003). *Public Policy and Public Participation Engaging Citizens and Community in the Development of Public Policy*. Retrieved Sept., 2010 from http://www.phac-aspc.gc.ca/canada/regions /atlantic/pdf/pub_policy_partic_e.pdf

UN E-Government Survey. (2008). *From E-Government to Connected Governance*. Accessed Jan, 2011 from http://unpan1.un.org/intradoc/groups/public/documents/un/unpan028607.pdf

Walker, M., Litman, D., Kamm, C., & Abella, A. (1997). PARADISE: A framework for evaluating spoken dialogue agents.*Proceedings of the 35th Annual Meeting of the Association of Computational Linguistics*. doi:10.3115/976909.979652

KEY TERMS AND DEFINITIONS

Electronic Democracy (E-Democracy): Anything that governments do to facilitate greater participation in government and enhance effective governance using digital or electronic means.

Electronic Participation (E-Participation): The use of internet and mobile technologies to enhance democratic governance.

Inclusive Political Participation: Government and citizens political interaction that does not exclude any particular groups of people.

Information and Communication Technology (ICT): Collection of devices, protocols, procedures and networks employed in the information technology fields to foster interaction among different stakeholders.

Participation: Taking part in joint activities for the purpose of reaching a common goal i.e. direct public participation in political activities.

Chapter 4
Embracing Biometric Voter Registration in Africa:
Pros and Cons

Emmanuel Eilu
Makerere University, Uganda

ABSTRACT

One of the major reasons for poor voter turn-up in transitional and post conflict countries in Africa is the credibility and legitimacy of the voter's register. Many voters are disfranchised, and many of the voter's registers are erroneous and lack credibility. This has created a wave of mistrust and discontentment amongst the voters and many prefer not to participate in the election process. Yet, the most fundamental form of exercising good democracy is through participating in regular free and fair elections. However, before a free and fair election is carried out, relevant electoral bodies and authorities normally establish a comprehensive, credible and inclusive voters' list, also called a voters' register. This chapter discusses manual voter registration in Africa and its challenges. It further discusses biometric voter registration systems in different African countries; the opportunities it provides to pacify and redeem voters' registers. Finally, this chapter informs relevant governments in Africa on the most critical success factors for adopting biometric voter registration.

INTRODUCTION

The post-colonial era in many African nations was characterized by various forms of dictatorship and autocracy, and many military coups and instability were rampant. Only a few countries namely; Botswana, Mauritius, and The Gambia experienced

DOI: 10.4018/978-1-5225-2565-3.ch004

extended periods of peaceful multiparty democratic elections (Burchard, 2015). However, by the early 1990s, a great wave of democratic elections swept through a number of African countries. According to Burchard (2015), the majority of nations in sub-Saharan Africa moved away from authoritarianism and moved towards democratic electoral regimes. However, some countries practiced anocracy, or hybrid regimes that occupy the space between democracy and autocracy (Burchard, 2014). In this kind of hybrid democracy, many regimes restrict political freedom and competition and use elections more as a façade for their continuing rule, as in the Central African Republic (CAR) and Rwanda (Vorrath, 2011). There was an increase in the number of anocratic regimes in Africa from 17 in 1991 to 24 in 2012 (Burchard, 2014). In fact, almost half of the 24 countries that have recently held elections are showing a worrying tendency to maintain or even step up authoritarian measures and political exclusion (Vorrath, 2011). However, 2000 to 2012 saw significant gains where many countries transitioned from anocracy to full democracy (Burchard, 2014). Compared to the authoritarian type of governance or even anocracy, full democracy has been fronted as the most preferred method of governance for a number of reasons. Frist, democracy is believed to be more conducive for economic growth and social development (Lipset, 1959; Burkart, 1995; Przeworski et al., 2000). For example, in nations such as Ghana, Mauritius, Namibia and South Africa, free and fair elections led to the emergence of strong democratic institutions and huge leaps in economic and political development (CAE, 2009). Secondly, democratic governance is also believed to be a more stable system of government in the long run compared with dictatorships, single-party systems, and the like, which are prone to civil strife and coups d'état (Feng, 1997; Przeworski et al., 2000). In Liberia and Sierra Leone for example, credible elections have fostered national reconciliation and an end to years of armed conflict and civil war (CAE, 2009; Burchard, 2015).There are very low chances that democracies engage in significant external conflict or war with each other, making the wide spread of democracy a likely precondition for global peace (Owen, 1994; Hensel et al., 2000).

Whereas Africa is celebrating wide-spread formal democratic reforms cultivated since the beginning of post-colonial era, there is still a lot to be done in deepening its democratic reforms. This is because the electoral processes in many African countries continue to be marred by accusations of bad practices, boycotts, gerrymandering, forms of repression and sheer violence (DDP, 2008). According to Vergne (2009), violence, intimidation and fraud has been recorded in most developing countries. Electoral fraud According to Systemic Peace report (2012) is a common occurrence and more than 50 percent of elections in Africa can be characterized by harassment, intimidation, or some other form of electoral malfeasance. Approximately 14% of all multi-party elections in developing countries were boycotted from 1990-2002 as a result of election violence, fraud and intimidation (Beaulieu & Hyde, 2009). It is also evident that Citizen's interest in democratic participations, and more particularly in general elections in many African countries is dwindling, and this has made many governments loose legitimacy. Africa as a continent has the lowest voter turnout (65%) in the entire world (Vergne, 2009). The voter turn-out in countries like Uganda under-score this assertion. For example, the 2005 referendum in Uganda had a 47% voter turnout (Petersen, 2006), while in the 2011 presidential elections registered 57% voter turn-out (EU, 2011; Oola, 2011).Tanzania's case is not different, in 1995 elections, there was a 76% voter turn up, but kept on falling through subsequent elections to the lowest of 43% in the 2010 elections. However, there was a slight improvement in 2015 with about 67% (Mungai, 2015; IIDEA, 2015). The 2015 elections in Nigeria had 43% voter turn up, down from 53% in 2010. This kind of trend is registered in many other countries like Mali, Zimbabwe and Burkina Faso among others had registered a voter turn up of less than 60% (IIDEA, 2015). When citizens do not turn up to vote, their willpower will not be demonstrated. They lose out on equal representation. According to Pintor et al., (Nd), voter turnout is measured as the extent to which citizens participate in how they want to be governed and by who. *It is usually expressed as the percentage of voters who cast a vote* (i.e., "turnout") *at an election* (Pintor *et al.,* Nd). By voting in competitive elections, for example, they hold within their authority, the power to grant political legitimacy on any leader or withhold it from any leader, institution and regime (Bratton, 2013).

However, besides the violence, fraud and low voter turn up witnessed in many African countries, one of the major challenges many governments in transitional and post conflict countries in Africa are grappling with is voter registration and the legitimacy of the voter's register. This has stretched democracy to its greatest limits in many of these countries. There is a lot of mistrust not only in the voter registration process but also in the voter's list. Many of the voter's registers are erroneous and lack credibility. When the citizenry realizes that there are enormous errors and ''ghost'' voters in the national voter's register, this create a wave of mistrust, disfranchisement, and discontentment amongst the voters and many will prefer not to participate in the election process (Nzalendo, 2012; EU/UNDP, 2010). According to Evrensel (2010), the role of voter registration is extremely significant when it comes to emerging democracies: this process can make or break an election. The quality of the process and of the the voters' roll determines the outcome of an election and consequently the stability of the democratic institutions in a country (Evrensel, 2010).

This chapter therefore, explores voter registration and the problems associated with the manual voter rolls in post conflict and emerging democracies in Africa. This chapter also highlights potential benefits and points out likely challenges of using technology, particularly biometric voter registration technology. The chapter finally outlines and discusses a number of critical success factors for adopting and implementing biometric voter registration system in post conflict and emerging democracies in Africa. This chapter informs relevant governments in Africa on the most critical success factors when adopting biometric voter registration.

VOTER REGISTRATION IN AFRICA

Democracy- according to Bratton & van-de Walle (1997) has become a pillar of contemporary political and development discourse in Africa and seems more pervasive than ever in the continent's history. The right of all voting age citizens to participate in the affairs of their government is one of the pillars of democracy (EKN, 2016). According to Pintor & Gratschew (2011), the right to vote in elections is universal, equal, direct and secret. This exercise is the means through which the governed agree to delegate their authority and power to those who govern. National leaders must be approved by the people, it is only when leaders fully obtain the mandate to govern from the people that they will be efficient, transparent and accountable to the electorate and the government. Because we cannot all be presidents at the same time, we will have to choose one person to take up the presidency on our behalf and he or she represents our views. This is normally referred to as participatory governance or democracy. The most fundamental form of exercising participatory democracy is through regular free and fair elections. However, before a free and fair election is carried out, relevant electoral bodies and authorities need to register voters (voter registration) and establish a comprehensive and inclusive voters' list, also called a voters' register (EKN, 2016). Voter registration is the first critical step in participatory democracy (Pintor & Gratschew, 2011). Voter registration can be defined as the process of registering eligible voters, while the voters' register or voters' roll is the result of this process (Evrensel, 2010;EKN, 2016). Bodnár & Kaszás, (2009) on the other hand defines voter registration as a process where the electoral management collects data of those individuals who are eligible to vote. Both the process and the result of voter registration need to be accurate, sustainable and politically accepted, and the list must be carefully maintained to ensure that each eligible citizen is registered to vote once and only once in an election (Evrensel, 2010; EKN, 2016). According to EKN (2016), a voters' list clearly defines and separates two of the most important functions of the election management authority, namely verifying voter eligibility and controlling the legitimacy of the voting process. Voter registration must be fair and inclusive in that the restoration exercise must be accessed by groups and categories of eligible citizens. In other words, the register should include every adult

Table 1. Periodic voter lists

	DRC	Ghana	Liberia	Malawi	Mozambique	Rwanda	Senegal
Population (million)	60	24	2.9	14.3	21.7	9.2	12
Registered voters (million)	25	12.8	1.35	5.9	9	4.7	4.9
Method for updating	Periodic – 2005, 2009	Periodic	Periodic – last 2005, next 2011	Periodic – updated before election, staggered process	Periodic – updated before elections	Continuous with periodic updates before each election	Periodic – every year

Evrensel, 2010.

person belonging to the citizenry (Pintor & Gratschew, 2011). There should be no systemic bias against any identifiable group. For registration to be fair, comprehensive, and inclusive, potential voters must be aware of the registration process and have reasonable opportunities and relatively easy access to complete it. Voter registration can either be voluntary or mandatory. In the case were voting is an obligation and responsibility of the citizen, then registration becomes mandatory as in the case in Australia or Indonesia. In that case, the state must make voters aware that they are required to register, and/or to update their information on the continuous voters' list. If voting is viewed as a right rather than an obligation, registration is an option to be exercised by voters at their own discretion as it is the case in the United States, Uganda, Guatemala and many other countries (EKN, 2016).

There are mainly two types of voter registers. In many cases, the election authority may have to choose one of these, or a combination, as the basis on which to design a voter registration system as discussed by (EU/UNDP, 2010; Pintor & Gratschew, 2011). A periodic register of voters, or a "periodic list," is established for a specific electoral event, and electoral administrators do not necessarily intend to maintain or update the list for future. The periodic lists may be appropriate for government especially in low developed economies without adequate infrastructure to maintain the voter list. For example, in Uganda, there has been voter registration prior to every presidential election since 1996. Other countries that use periodic voter lists in Africa are presented in the Table 1.

A continuous list on the other hand is a voter list compiled, maintained and regularly updated by the electoral administration, such as the continuous voter register used by the South African Election Commission. This system depends on an appropriate infrastructure to maintain the list. Maintenance typically involves performing a lot of audit work such as; adding the names and other relevant information of new eligible voters, deleting the names of those who no longer meet the requirements (e.g. through death); and altering the details of those voters whose recorded data in the voter list requires updating e.g. through name change from marriage or change of residence (EU/UNDP, 2010; Pintor and Gratschew, 2011)

According to Rosenbergwith and Brennan (2009), there are two ways in which the electoral body can generate voters' list. First, the voters list can be generated through mass registration at designated places, where the eligible individuals must initiate their own voter registrations in-person at a local election office or a designated place, such as in the Bahamas, Burundi, Uganda, Belize and many other countries. Alternatively, the government could conduct a nationwide enumeration by hiring temporary workers to go door-to-door, registering every eligible voter such as in Indonesia and Canada Saskatchewan. Secondly, election officials may use information collected from other government agencies to either come up with a new voters list or add eligible citizens to the already available voters rolls, such as in France, Germany, Australia and Belgium to mention but a few (Rosenbergwith and Brennan, 2009). However, EU/UNDP (2010) raises concern on the use of civil registers for compiling voter list and observes that- data sharing among government institutions may be controversial. Concerns may be expressed about the loss, or potential loss, of privacy - particularly if the government decides to add additional data fields or merge the registry with, for example, the tax authority's database. If the concerns are widespread, a civil registry may be simply unacceptable despite its usefulness. In the next section, this chapter identifies and discusses the two major types if voter registers.

Types of Voter Registers in AFRICA

Low-Tech Electronic Registers

Many countries in the developing world and especially Africa are still using *low-tech* electronic registers, these are electronic registers compiled by manual entry of data from paper forms compiled at the voter registration centers. This was often chosen in "first generation" or post-conflict elections when technology was not available or was far too expensive (EU/UNDP, 2010). The manual voter registration in many cases is extremely erroneous. The margin for error or manipulation in the manual system of voter registration is great (Nzalendo, 2012): As stated before, the *low tech* electronic register does not have an in-built mechanism for detecting multiple registrations. The voters' lists in which manual paper registration was used for the past elections in 2007 and 2011 in Kenya, contained a large number of duplicate or invalid registrations. The election official in some cases mistakenly crossed out the voter's names, wrongly entered information on the voter's card, and incorrectly entered information on the voter registration form (Nzalendo, 2012). Likewise in Uganda, the manual voter registry contained a lot of errors during the 2006 elections that the Electoral Commission confessed deleting over 150,000 'ghost voters' from the register at the conclusion of 2006 elections (HURINET-U, 2010). While in 2010, the same Electoral commission in Uganda began identifying and deleting duplicate entries in the printed manual register, as a result, over one million duplicate voter entries were identified and deleted (EU, 2011). Therefore, the legitimacy of the entire process will instantly be discredited if there are problems with the registration of voters, and particularly with the integrity of the voters' list (Nzalendo, 2012; EU/UNDP, 2010). There is a wider call for governments in Africa to have clean voter registers free from "Ghost voters".

Biometric Voter Registers (BVR)

In response to a wider call for governments in Africa to have clean voter registers free from "Ghost voters", more democratic countries are increasingly adopting modern, efficient, effective and secure voter identification technologies to help support credible and transparent elections (Shahnewaz, 2014). As information and communication technology (ICT) trends advance and challenges continue to emerge for manual voter registration, many developing governments such as Democratic Republic of Congo (DRC), Togo, Guinea Conakry, Uganda, Angola, Nigeria and Mozambique have taken the wider strides to replace their existing voter register with a biometric register to ensure the effectiveness of the "one person one vote" (Caarls, 2010; Shahnewaz, 2014). Biometric voter registration (BVR) has recently emerged as a new way to register voters, and Election Management Bodies (EMBs) across the African continent have increasingly turned to biometric technologies in order to address the challenges of managing their voter registries (Shahnewaz, 2014). The quest for biometric identification systems in Africa has been steadfastly growing since the early 2000s. Biometric technology has frequently been preferred as an effective way to avoid disputes over the voters' roll and to minimize electoral fraud. Biometric voter registration (BVR) systems store one or more physical characteristics of the voter such as; Automatic Fingerprint Identification (AFI) which normally requires good reliable fingerprints to be place and later collected through a high definition scanner, and then Facial Recognition which requires clear faces to be photographed. The goal is to prevent duplicate entries on a voters list, and more accurately confirming a voter's identity on Election Day (Shahnewaz, 2014; Mathieson, 2011). The system ensures that a voter is who he/she claims to be and only allows to vote one time. The biometric voter registration decreases the inefficacies of voter registration and has a high ability to eliminate "ghost" voters (Nzalendo, 2012). According to EU/UNDP (2010), biometric systems have proven

Figure 1. A computerized voter registration kit
Evrensel, 2010.

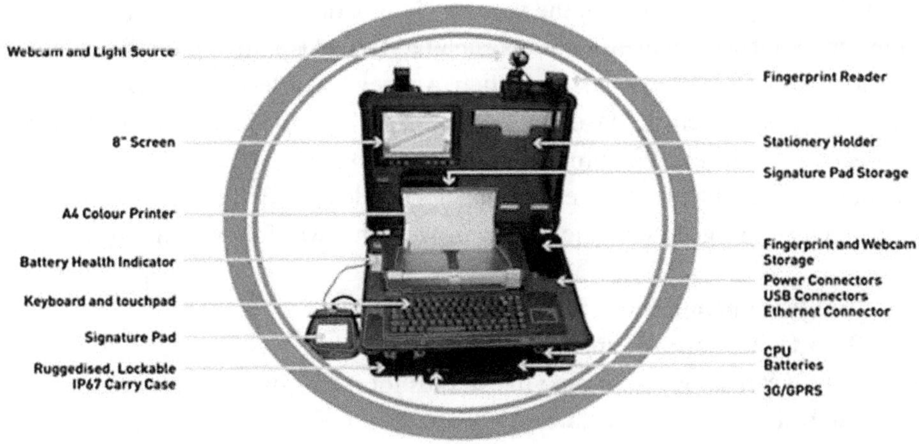

particularly useful in two specific segments of the electoral cycle, namely voter registration and voting operations. If AFI technology is used during voter registration, duplicate registrations can be detected. Biometric systems are in fact increasingly considered to be the definitive solution to voter impersonation and multiple voting practices. A sample of the biometric voter registration kit used in many African countries is presented in Figure 1.

Biometric voter registration has always presented an opportunity to many developing countries to pacify and redeem their erroneous manual voter lists. In the next section, this chapter discusses the opportunities of embracing biometric voter registration.

Methodology: Systematic Review

Systematic review will be the main method used in this chapter. In the last 30 years or so, narrative-style literature reviews have been criticized as being biased and inadequate in terms of the rigor of research performed (Hemingway & Brereton, 2009). This has led to the emergence and wide spread use of systematic review method of research. Systematic Review is a 'rigorous method used to map the evidence base on an unbiased way as possible, and to assess the quality of the evidence and synthesize it' (DFID, 2013). Systematic review method follows a formal process for appraising literature and minimizing bias (Kowalczyk & Truluck, 2013). According to Zanker and Mallett (2013), Systematic review is considered by some to offer 'the most reliable and comprehensive statement about what works and how it worked. It has been widely used for many years in different fields such as medical research and the natural sciences. Systematic review has many times been employed in evidence-informed policymaking within the arena of international development. It is much preferred by international agencies as such the UK's Department for International Development (DFID) and the Australian Agency for International Development (AusAID). These agencies have funded a series of Systematic reviews over the past few years, with the express aim of finding out 'what works and how it worked' in generating development outcomes to (Zanker & Mallett, 2013). In 1984 Cooper (1984) proposed a five stage systematic review process that was followed in this chapter'

- **Problem Formulation:** Statement of objective.
- **Data Collection:** An unbiased literature search.
- **Data Evaluation:** Assessing the studies for inclusion in the review.
- **Analysis and Interpretation:** Qualitative or quantitative aggregation of individual research studies.
- **Public Presentation:** Discussion and context of findings.

According to Kowalczyk & Truluck (2013), systematic review is considered original research as it follows a standard scientific protocol.

OPPORTUNITIES OF BVR

Reduced Duplication

With biometric registration, detecting and removing of duplicate registrations, correcting wrong entries and performing audit can relatively be easy. Continuous registration for voters who attain the age of 18 years is easy, even after the general elections. As a result, a more accurate, error free and reliable voter register can be produced to support free and fair elections. For example, in 2005 in Liberia, a biometric voter register was displayed for public scrutiny for a three day period. Only about 0.6% of all registered voters' data had errors, and these were all corrected. There was no contestation of voters' register among the competing political parties. While in DRC, duplication was not regarded as a key threat to the exercise, de-duplication did not score much following an under 1% duplicate detection rate in the biometric voter registration exercise (Evrensel, 2010).

Improves Citizen Participation in Governance

Biometric registers have gained citizen trust and credibility in many democracies and therefore have led to improved citizen participation in political governance. Empirical evidence from emerging democracies shows that whenever comprehensive biometric voter registers are created, or existing registers of a limited scope are substantially improved, voter registration and voter turnout rates tend to be, on average, as high as in most established democracies in Europe (Pintor and Gratschew, 2011). For example, in Democratic Republic of Congo(DRC), over 80% of the 25 million registered voters turn up to vote as a result of credible biometric registration system exercise carried out prior to the 2006 elections (IDEA 2014; Evrensel, 2010). Sierra Leone on the other hand introduced a biometric voter registration system, held peaceful, free and fair elections on November 17 2012, with 87% voter turnout (UNDP, 2012). Over 90% of the eligible voters were successfully registered in Malawi, DRC, Ghana and Liberia (Evrensel, 2010). Nigeria, which saw one of the most remarkable swings in Africa's democratic history in 2015 general elections. The Nigerian government significantly revised their pervious voters' register with a new biometric voter registration system. Permanent voter cards with a biometric chip of the voter's unique details were issued out. This new and effective biometric registration system cleaned up the national voter register, reducing Nigeria's voter population from 74 million in 2011 to 68 million in 2015, contributing to the cleanest and peaceful election in Nigeria's recent history (Quaicoe, 2015)

Easy Settling of Disputes

Another major benefit of undertaking biometric voter registration before the election period is that, it allows disputes about the right of a person to vote to be dealt with in a measured way, well before the polling takes place, and thereby minimizes disputes on Election Day (Pintor & Gratschew, 2011). In 2010 after the biometric voter registration process and before the 2011 presidential election, the Electoral Commission in Uganda issued electronic copies of the voter register to political parties and their presidential candidates, and on 31 January 2011 it provided printed copies including photographs for verification. A few errors were pointed out and they were immediately corrected easily before the Election Day (EU-2011).

It's Faster and Has Synergy Effect

The process of voter registration is faster with biometric system than in a manual system. In just 6 weeks, 2.7 million voters were registered in Sierra Leone, while Ghana successfully registered 14 Million potential voters within 40 days. Interoperability between government agencies for database consolidation and maintenance becomes easier when biometric voter registration is done. According to Evrensel (2010), the synergy effect is great in that a successful combined civil and voter registration process is less costly and produces shared databases that have multiple uses by different government departments and agencies. In the case of Rwanda, the civil and voter databases were established from scratch in a combined effort. Field registration teams collected information for both databases. The process was efficient as equipment was shared and there was no duplication of work and tasks. In Malawi, a high-integrity voter ID card has been accepted widely as a generic national identity card (Evrensel, 2010).

Whereas there are enormous advantages of biometric voter registration technology, however, every technology comes with its challenges and biometric voter registration systems are not exceptional. The next section highlights some of the challenges of BVR

Challenges of Using BVR

Evrensel (2010) observes that a complex voter registration system does not guarantee successful elections. On the contrary, the use of sophisticated technology can be a threat to the acceptance, transparency and security of the system. Some of the particular challenges associated with the introduction of voter registration biometric technology are highlighted from Uganda's experience of 2000 prior to the 2001 presidential: The Ugandan project, also known as the voter registration database implementation system, was a failure prior to the 2001 elections. The government of Uganda invested $22 million on equipment, consultancy services and operations. The project failed as a result of numerous challenges like; time required to register voters was not adequate; un trained/skilled data capture and maintenance staff, procurement challenges, there was a lot of opposition from other political parties, lack of political will, inadequate and unfair ICT laws, misconception by some people that the biometric registration kits were harmful to the body to mention but a few. Even in Ghana's last election in December 2012, which is considered best practice for many African countries encountered a number of glitches in the biometric voter register. The biometric kits failed in many parts of the country, forcing voting to be extended to the second day. In the 2013 elections in Kenya, both the biometric kit and the electronic tallying system malfunctioned, this force the electoral authority to manually tally vote. This was the same case with the semi-autonomous region of Somaliland in 2010. Many failures have also been reported in such projects in many developing countries; such as in Nigeria in 2007, Peru in 2000, Nicaragua in 2008 and Cameroon 2012 (Heeks 2003; Blanc 2007; Lopez 2011; Wrong 2013; Ogot 2014; Mungai, 2015).

Furthermore, this new technology does not come cheap, and only donor support makes this biometric registration exercise possible (Lopez 2011; Wrong 2013; Ogot 2014). The entire task of registering voters and producing voters' lists often accounts for close to 50 percent of the overall cost of administering elections (Evrensel, 2010). For example, In DRC, the recent elections cost a colossal $360 million, with $58 million of that spent on biometrics voter registration system. While in Ghana, the figures were $124 million and $76 million respectively. Kenya's elections cost $293 million, with donors putting in $100 million in the biometric system (Wrong, 2013; Mungai, 2015). Uganda spent $20 million in a failed e-registration system in 2001 (Heeks 2003). While Malawi spent about $43 million on purchasing a biometric system. Tanzania has already budgeted about $50 million to purchase equipment for the 2015 general elections (Ogot, 2014).

In summary, it is a fact that biometric voter registration is difficult to implement, specification and procurement processes are lengthy, training of operators and data processing is a complex and lengthy process which puts a lot of pressure on the Electoral Commissions to produce the list in time. Some countries eventually fail to implement the biometric verification of the voter which renders the entire exercise null and void, therefore losing billions of dollars. However, a number of strategies can be undertaken for a successful implementation of biometric voter registration. The next section discusses some of the strategies that can be undertaken by the government for successful BVR processes

Success Factors for BVR

Effective Campaign

There is need to popularize the biometric voter registration system thorough an effective awareness campaign. Governments through the National Commissions for Civic Education (NCCE) needs to embark on a nationwide campaign to intensify their education programmes on the biometric registration process. Thorough and effective outreach campaign on the biometric registration system is needed to ensure eligible voters have their names on the biometric register and they are not disenfranchised. Among the widely used channels for sensitization include; TV, Radio, Newspaper adverts and many others. In the modern political age, technology is the backbone of any campaign; well designed tool will allow the NCCE to share information cheaply, easily and efficiently. Tools like twitter, facebook and many other social media platforms could turn out to be the most effective tools for sensitization. For example, the success of biometric voter registration process in Ghana is partly linked to the creation of *Ghana Decides*, a non-partisan project before the elections of 2012 utilized various social media sites such as Storify, Google+, Youtube, Facebook, Flickr and Twitter to reach voters. They also work to dispel common concerns that BVR causes cancer and promote the technique as a safe and accessible process.

Pilot Scheme

Piloting is one of the standard methods for measuring the success or failure of a project. A pilot experiment is a small scale preliminary study conducted in one or a few selected areas in order to evaluate feasibility, time, cost, adverse events, and effect size of a given project. There is need for National Election Bodies to carry out pilot experiments so as to determine the projects likelihood of success or failure before the final full implementation. For example in Ghana, Ashanti region was used to pilot the biometric voter registration before it was later spread across the whole nation. Ghana's biometric voter registration in 2012 was widely viewed as a success. In Sierra Leone, after a successful pilot project of District Code of Conduct Monitoring Committees (DCMCs) in which party representatives, civil society and other stakeholders were trained in conflict mediation, prevention and resolution, the government rolled out a national programme in all 14 districts, helping to diffuse tensions and acting as an early warning mechanism in the run-up to elections and also refine operational procedures and software for the roll-out of the system in 2012.

Political Support, Strong Institutions, and Fair Laws

The success or failure of e-government projects - especially those involving citizens and the democratic process - is significantly determined by the willingness of the executive to support the initiative. Unless there is a political will to see the biometric system succeed, it is likely to fail. The success of biometric voter registration exercise is also strongly linked to strong, accountable, independent and transparent institutions especially the Legislature, Judiciary and the NCCE. Institutions such as representation systems and consultation committees ensure that policymakers must consult with the electorate and other elected officials before making policies. Without these institutions and practices, elected leader may act like dictators, enacting and implementing policies which are in their interest. Birch (2008) observes that -in manipulating the election rules, the strategy carries a fairly low level of risk, rules of the game can be changed through legal means, and though there are international legal standards governing any electoral process, state actors nonetheless benefit from considerable leeway in the interpretation of international norms. Lopez (2009) refers to manipulation of rules as corruption, arguing that- it is a self governing demeanor that deviates from formal guiding principles governing someone in a position of public authority. In Uganda for example, the Law allows the seating president to appoint the chair person of the Electoral Commission (EC) and all the 8 commissioners. This Act has always been objected by other political parties. The opposition parties see the EC chairperson and thc 8 commissioners as partisan and lean towards the seating president who appointed them. The Electoral Laws and Electoral Commission should gain acceptance, trust and prove its fairness by addressing the concerns of all stakeholders in a transparent manner to ensure zero grievances from parties. Electoral Commission should put in place measures that will allay the fears of all political parties in the forthcoming nationwide biometric registration exercise. ICT equipment, including hardware and software, needs to be legally supported and/or adapted to the specific legal context of the country

An All Inclusive Process

To register any success in the biometric registration exercise, the government should ensure active participation of all stake holders such as civil society, the diplomatic community, the media, all political parties, religious groups, women, youth and many others. The Sierra Leonean government for example with support from UNDP ensured that international standards and best practices, and views and concerns from all electoral stakeholders through consultative processes, informed the drafting of the Public Elections Law and the Political Parties Registration Commission (PPRC) Act. Prior to the general elections in 2012 in Sierra Leone, all 10 registered political parties and other major stakeholders had signed a declaration before the registration exercise, committing themselves to free, fair and peaceful elections, and since then, there had been no incidents of political violence and commitments were being upheld. The media also had an important contribution to make, ensuring a free, fair and peaceful election by providing objective and balanced information. In close cooperation with other organizations, the All Political Parties Women Association (APPWA) embarked on increasing women's participation in political processes and the All Political Parties Youth Association (APPYA) helped to diffuse tensions through the sensitization of voters on peace and tolerance in Sierra Leone.

Efficient, Effective, and Secure Biometric System

An efficient biometric voter registration systems should be able to collect, transmit, compile, clean, audit and store voters' data in a fast, efficient, effective and secure manner. The appropriateness of the registration system will depend on what is expected of the system and the efficiency of the planning and operations involved.

Social-culturally accepted technology should be taken in to account. In Ghana for example, the EC procured scanners that are battery operated and can work for 16 hours a day. In any particular case, the registration is 11 hours a day. In Sierra Leone, UNDP provided extensive technical support to NEC in the design, procurement and piloting an efficient and effective Biometric Voter Registration (BVR) which was the first of its kind in the country. In general terms, the BVR system must have all the hardware and software needed to identify eligible voters and issue them with either the voter cards or proof of registration on the spot.

Training Election Staff and Technical Assistance

Recruiting qualified personnel and conducting proper training for electoral staff and presiding officers is a major factor in ensuring a successful biometric registration process. In DR Congo, Zetes-a hi tech company supplied 10,000 registration kits, and provided thorough training to the NCCE staff on the equipment and gave technical assistance throughout the registration period. In Sierra Leone, some 42,000 field personnel were trained in conflict resolution and violence related challenges, while 170 National Electoral Commission staff and commissioners were trained in the biometric registration technology, enabling them to register 2.7 million voters as well as successfully print and distribute voter identity cards across the country in the run-up to the 2012 elections. While in Liberia, 4000 Liberians were hired and trained to work in the field for the voter registration process. Data was collected in the field on pre-printed optical mark recognition (OMR) forms, and voter cards featuring a black and white photograph of the voter were produced using Polaroid cameras

Funding

Undertaking a biometric voter registration processes is a very expensive venture for the world's poorest economies especially in Africa, and it counts for over 50 percent of the overall cost of entire election process. As a result of meager resources, a greater percentage of these biometric registration systems in Africa have been externally financed. There is need to seek funding for the biometric voter registration exercise from the international community. In Nigeria, the United States is providing approximately \$51 million over five years to support the conduct of credible and peaceful elections in 2015 and beyond. In 2005, donors approved an estimated budget of some US\$285 million to support the electoral process in the DRC, and of recent the U.S. government announced \$30 million in additional funding for elections and stabilization, and has committed to working with the Congolese people, the government, and other donors to encourage the conduct of local, provincial, and national elections that are credible, timely, and in line with the Congolese constitution. A total contribution of US$18.9 million was given to the Liberian government by UNDP for the 2005 Liberian elections

CONCLUSION

The most fundamental form of exercising participatory democracy is through regular free and fair elections. However, before a free and fair election is carried out, relevant electoral bodies and authorities normally establish a comprehensive and inclusive voters' list, also called a voters' register which is free from errors. However, voter registers and registration systems used by a number of developing countries especially in Africa are erroneous, slow and outdated. A flawed voter registry complicates electoral planning and removes an important safeguard against fraud. In an attempt to have clean voter registries that underpins credible elections, a number of African countries have dived in to wave of the ''magic bullet'' known as biometric voter registration system. Biometric voter registration when well implemented has a potential of significally reducing or eliminating errors and ghost voters in the national voters register. It is good verification and foolproof method of identifying each voter and in many cases delivered a free and fair electoral process. Without doubt, just like any other technology, a biometric voter registration process and voter registers have their own challenges, ranging from social, political, technical and financial challenges. However, these challenges can be addressed through designing and following a well thought out implementation process such as; recruiting and thoroughly training election staff, securing funding, procuring an effective and efficient system, stake holder involvement and many others as already discussed in this chapter

REFERENCES

Ake, C. (2000). *The Feasibility of Democracy in Africa*. Antony Rowe Ltd.

Beaulieu, E., & Hyde, S.D. (2009). *In the Shadow of Democracy Promotion: Strategic Manipulation, International Observers, and Election Boycotts*. Department of Political Science, University of Kentuck7.

Birch, S. (2008). *Patterns of Electoral Malpractice: Project on Electoral Malpractice in New and Semi-Democracies*. Working Paper No 4. Department of Government, University of Essex. Accessed on 08th Jan 2016 http://www.essex.ac.uk/government/electoralmalpractice/

Blanc, J. (2007). *Challenging the Norms and Standards of Election Administration: Electronic Voting*. Accessed January 12, 2016, from http://www.ifes.org/publication/de870bce4b39339d89a3469c8de3a0a0/1%20IFES%20Challenging%20Election%20Norms%20and%20Standards%20WP%20ELVOT.pdf

Bodnár, E., & Kaszás, A. (2009). *Theory and Practice of Voter Registration – Definitions, Standards, Principles, Examples*. Association of European Election Officials.

Bratton, M., & van de Walle, N. (1997). *Democratic Experiments in Africa: Regime transitions in comparative perspective*. Cambridge University Press. doi:10.1017/CBO9781139174657

Burchard, S. M. (2014). *Democracy Trends in Sub-Saharan Africa, 1990 to 2014*. IDA Document NS D-5393.

Burchard, S. M. (2015). Electoral Violence in Sub-Saharan Africa: Causes and Consequences. Academic Press.

Burkart, R. E., & Black, M. S. L. (1995). Comparative Democracy: the Economic Development Thesis. *American Political Science Review, 88*(4), 903–910.

Caarls, S. (2010). *The Use Of Technologies In Electoral Systems Prepared For IFES-Egypt.* Retrieved January 12, 2015 from http://www.aswatna-eg.net/assets/files/Technology%20and%20Elections.pdf

CAE -Colloquium on African Elections. (2009). *Best Practices and Cross-Sectoral Collaboration Final Report.* Accessed 15 December 2015 on http://www.ndi.org/files/African_Elections_Best_Practices_ENG.pdf

DDP-Democracy Development Programme. (2008). Participatory Democratic Governance In Africa: From A 'Democracy Of Voters' To A 'Democracy Of Citizens'. Paper prepared by the Research Unit for the 'Good Governance Africa sector network' conference, Addis Ababa, Ethiopia.

DFID-Department for International Development. (2013). *Systematic Reviews in international Development*. Accessed on 10th March 2016 from https://www.gov.uk/government/publications/systematic-reviews-in internationaldevelopment/systematic-reviews-in-international-development

EKN-Electoral Knowledge Network. (2016). *Voter Registration*. Retrieved in March 24th 2015 from http://aceproject.org/ace-en/topics/vr/default

EU-Election Observation Mission. (2011). *Uganda, general elections, preliminary statement uganda 2011 elections: Improvements marred by avoidable failures*. Author.

EU/UNDP-European Union/United Nations Development Program. (2010). *Procurement Aspects of Introducing ICT Solutions in Electoral Processes*. The Specific Case of Voter Registration.

Evrensel, A. (2010). Voter registration in Africa. A comparative analysis. EISA.

Feng, Y. (1997). Democracy, Political Stability and Economic Growth. *British Journal of Political Science, 27*(3), 391–418.

Heeks, R. (2003). *IDPM Most e-government-for-Development Projects Fail How Can Risks be Reduced?* Retrieved January 19, 2015 from http://unpan1.un.org/intradoc/groups/public/documents/cafrad/unpan011226.pdf

Hemingway, P., & Brereton, N. (2009). *What is a systematic review?* Hayward Group Web site. Accessed on 08th Jan 2016 from http://www.whatisseries.co.uk/whatis/pdfs/What_is_syst_rev.pdf

Hensel, P.R., Goertz, G., & Diehl, P. F. (2000). The Democratic Peace and Rivalries. *Journal of Politics, 62*(4), 1173–88.

HURINET-U-Human Rights Network Uganda (2010). *Towards peaceful, free, fair and Democratic 2011 elections in Uganda: An analysis of Uganda's electoral laws*. Author.

International Peace Institute. (2011). *Elections and Stability in West Africa*. Retrieved in February 2nd 2015 from http://reliefweb.int/sites/reliefweb.int/files/resources/Full_Docu_2.pdf

Kowalczyk, N., & Truluck, C. (2013, November/December). Literature Reviews and Systematic Reviews: What Is the Difference? *Radiologic Technology, 85*(2). PMID:24255144

Lipset, S. M. (1959). Some Social Requisites of Democracy: Economic and Political Development. *American Political Science Review, 53*(1), 69–105.

Lopez, P. R. (2011). *Assessing electoral fraud in new democracies: A basic framework*. What Happened, Sao Paulo, Brazil: IPSA-ECPR Joint Conference.

Breuer, F. (2006). *E-voting: Lessons learnt and future challenges conference report*. Academic Press.

Mathieson, D. (2011). *ICT in the Electoral Process in Africa*. MA MBCS CITP. Retrieved in March 24th 2015 from http://www.cmd-kenya.org/files/ICT_in_the_Electoral_Process_in_Africa_by%20David_Mathieson.pdf

Mungai, C. (2015). *Dirty hands: Why biometric voting fails in Africa - and it doesn't matter in the end.* Retrieved in May 5th 2016 http://mgafrica.com/article/2015-03-30-why-biometric-voting-fails-in-africa-and-why-it-doesnt-matter

Nzalendo. (2012). *Biometric Voter Registration vs. Manual Registration.* Retrieved January 12, 2015 from http://www.mzalendo.com/blog/2012/08/06/biometric-voter-registration-vs-manual-registration/

Quaicoe, N. A. (2016). *Voter Registration Can Make or Break an Election.* Accessed in June 12th 2016 on http://www.danquahinstitute.org/index.php/1084-voter-registration-can-make-or-break-an-election-by-nana-attobrah-quaicoe-di

Ogot, G. (2014). *Use of technology in African elections not foolproof technology-in-African-elections-not-foolproof.* Retrieved January 19, 2015 from http://www.standardmedia.co.ke/m/?articleID=2000124434&story_title=Use-of-

Oola, O. (2011). *Insight on Conflict: Uganda's 2011 Election: Any Lessons For Local Local Peace Builder.* Retrieved from http://www.insightonconflict.org/2011/03/uganda-2011-election-lesson-for-local-peacebuilders/

Owen, J.M. (1994). How Liberalism Produces Democratic Peace. *International Security, 19*(2), 87–125.

Petersen, C. E. (2006). *Uganda: Presidential, Parliamentary and Local Council Elections 2006* [pdf]. University of Oslo. Available at: www.jus.uio.no/smr/english/about/programmes/nordem/publicati ons/nordem-report/2006/0806.pdf

Pintor, R. L., Gratschew, M., & Sullivan, K. (n.d.). *Voter Turnout Rates from a Comparative Perspective.* Accessed 15 December 2014 http://www.idea.int/publications/vt/upload/Voter%20turnout.pdf

Pintor, R. L., & Gratschew, M. (2011). *Voter Registration and Inclusive Democracy: Analysing Registration Practices Worldwide.* Retrieved January 19, 2015 from http://www.idea.int/publications/vt/upload/Voter%20registration.pdf

Przeworski, A., Alvarez, M. E., Cheibub, J. A., & Limongi, F. (2000). *Democracy and Development.* New York: Cambridge University Press. doi:10.1017/CBO9780511804946

Rosenbergwith, J.S., & Brennan, M.C. (2009). *Expanding Democracy: Voter Registration Around The World.* Center for Justice at New York University School of Law.

Peace, S. (2012). *Sub-Saharan Africa: Regimes by Type.* Accessed in July 20th 2016 from http://www.systemicpeace.org/polity/ssafrica2.htm

UNDP-United Nations Development Programme. (2012). *Support to the Electoral Cycle in Sierra Leone*. Retrieved in February 2nd 2015 from http://www.sl.undp.org/content/sierraleone/en/home/operations/projects/democratic_governance/support-to-the-electoral-cycle-in-sierra-leone-.html

Vergne, C. (2009). *Turnout in Developing Countries: The Effect of Mass Media on National Voter Participation*. Retrieved in January 12, 2015 from http://papers.ssrn.com/sol3/papers.cfm?abstract_id=1692347

Vorrath, J. (2011). *African Developments: Political Trends in Recent Elections in Sub- Saharan Africa*. Briefing Paper.

Wrong, M. (2013). *The Spectator; Africa's election aid fiasco*. Accessed January 12, 2016 from http://www.spectator.co.uk/features/8890471/the-technological-fix/

Zanker, J. H., & Mallett, R. (2013). *How to do a rigorous, evidence focused literature review in international development*. A Guidance Note. Accessed on 10th March 2016 from https://www.odi.org/sites/odi.org.uk/files/odiassets/publications-opinion-files/8572.pdf

ADDITIONAL READING

Piccolino, G. (2015). Making democracy legible? Voter registration and the permanent electronic electoral list in Benin. *Development and Change*, 46(2), 269-292. 2.

Adeosun, O. O., Ibitoye, A. O., & Alabi, J. O. (2015). Real Time E-Biotechnology Voting System; Using Secret Splitting. *International Journal of Electronics Communication and Computer Engineering*, 6(6), 624. 3.

Ashworth, S. (2012). Electoral accountability: recent theoretical and empirical work. *Annual Review of Political Science*, 15, 183-201. 4.

Ellerman, D. (2015). The democratic worker-owned firm: A new model for the East and West.

Bhattacharya, S., Duffy, J., & Kim, S. T. (2015). Voting with Endogenous Information Acquisition: Theory and Evidence (No. 151602).

KEY TERMS AND DEFINITIONS

Biometric Voter Registration: Use of technology to capture and store one or more physical characteristics of a person such as fingerprints, palm prints, facial recognition, Irish scan etc using a high definition scanner or camera.

Low-Tech Voter Register: An electronic registers compiled by manual entry of data from paper forms compiled at the voter registration centers.

National Election Bodies/Electoral Commission: An agency mandated by the constitution to register eligible voters and to conduct a free and fair election.

Pilot Scheme: A small scale preliminary study conducted in one or a few selected areas in order to evaluate feasibility, time, cost, adverse events, and affect size of a given project.

Voter Register: A compiled list of eligible voters.

Section 2

Using Information Technology to Support Secondary and Tertiary Education

Chapter 5
E-Learning Adoption among Selected Secondary Schools of Ogun State, Nigeria

Adedeji Oluwaseun Adewusi
Olabisi Onabanjo University, Nigeria

ABSTRACT

This chapter examines E-learning adoption in secondary schools of Ogun state. Constructivism Learning Theory was used to give theoretical explanation to issues of concern. Quantitative data was obtained from respondents who were purposively selected from 5 private and government owned secondary schools each. Results were analysed using both descriptive and inferential statistics. This chapter was revealing as it concluded that despite the roles ICT-driven instructional aids has been playing in education, schools in Ogun state, especially the government-owned ones, are yet to extensively avail and adopt them for teaching and learning processes as a result of certain challenges. Owing to this, E-learning facilities are to be made available and teachers should be exposed to the use of E-learning in teaching and learning through training and continuous training as to keep them abreast of the innovation in education which will also help boost the quality of students being produced.

INTRODUCTION

Education unlashes all human potentials and a critical factor to human life. It is generally accepted worldwide that education is the instrument "par excellent" for bringing about national development. Such development can be seen in the positive roles education play in the socio-cultural, political, economic and human aspect of

DOI: 10.4018/978-1-5225-2565-3.ch005

nation building (Olorundare, 2003). When appropriately implemented therefore, education has the capability of boosting the welfare and productivity of the citizens. Hence, it is not surprising that most government in developed and developing countries spend a sizable portion of their resources for the development of education. Nigeria's realization of the pivotal role of education can be seen with a perusal of the National Policy on Education (2013) document. For this to be achieved, the source and channel through which knowledge is passed to pupils/students cannot be over emphasized. This is to say that our educational products are not only determined by the quality of curriculum, but also by the quality of teachers and medium used in imparting knowledge on the students.

The world of technology is continually growing and it is necessary to state here that Information and Communication Technologies (ICT's) have fused all nations into one and has thus, made the world a global village. The fields of computer science, computer engineering, software development and others related to ICT have offered unlimited channels for learning through electronic platforms (Rozina, 2002). This was buttressed by Abimbade (2002) who argued that educational technology has expanded the width of education and created new ones, especially at the beginning of the 21st millennium. One of these new and rapidly expanding boundaries is E-learning which is offering tremendous advantage to education sector. E-learning as an aspect of Information and communication Technology (ICT) is relatively new in Nigeria's educational system. It has become a novel epitome in library services as well as educational sector with a mission to serve as a development platform for knowledge-based society. It is a departure from the conventional approach in curriculum implementation. E-learning has the potential to revolutionize teaching and learning (DfES, 2003).

Experts maintained that students who use ICTs gain deeper understanding of complex topics and concepts and are more likely to recall information and use it to solve problems outside the classroom (Apple Computer, 2002; Arinze, 2012). Therefore, the relevance of ICT to learning reaffirms that students must have variety of skills to express themselves not only through outdated channel of paper and pen but through digital systems such as audio, video, and animation designed software. It is observed from the aforementioned statement that there is a need for students to develop learning skills that will enable them to think critically, analyze information, communicate, solve problem and compete successfully with their colleagues all over the world. Against this background, the study examines E-learning adoption in Ogun state, specifically secondary schools in Ago-Iwoye, which comprises of both privately and publicly owned schools. Specifically, the study: compared the E-learning facilities available for teaching; examined the level of teachers' application of E-learning facilities in teaching; and identified the challenges facing the adoption of E-learning among selected secondary schools in Ago-Iwoye, Ogun State. Hence, the study hypotheses are:

H1: There will be significant difference between E-learning facilities available in private and public secondary schools

H2: There will be a significant relationship between availability of E-learning facilities and teachers application

H3: There will be significant relationship between challenges of E-learning adoption and its application

REVIEW OF RELATED STUDIES

E-learning means learning done through electronic platform(s). It is a digitalized type of knowledge acquisition in which texts, audio or sound, pictures, images, graphics and videos can be simultaneously presented to students. Learning through electronic means the learner, whether far or near has easy access to quality learning platforms. This can be linked to the fact that learner has robust and unlimited interactions with instructional contents, facilitators and other learners and are given support and appropriate time to make reasonable and responsible contributions to the learning process (Ajayi, 2005).

Several authors, experts and academicians have given various empirical insights into the present state of E-learning adoption in Nigeria. For instance, some submitted that E-learning is challenged with the problem of material devices such as fax and wireless applications, digital library, computer, computer laboratories, Internet and e-mail facilities, videophone systems and teleconferencing devices, digital classrooms, multimedia systems and the problem of multimedia courseware development among others (Effiong, 2005; Global Information Technology Report, 2005; Jegede & Owolabi, 2008; Nwana, 2012; Seiden, 2000; Uhaegbu, 2001). This was reaffirmed by Aboderin & Kumuyi (2013), who stressed that there is shortage of E-learning tools such as Internet, computers, email facilities, multimedia, scanner, printer, VCD player, and digital camera in Nigerian secondary schools.

Recent studies also show that ICT technical support and Internet facility are lacking in secondary schools. Furthermore, teacher's expertise and lack of knowledge to evaluate the use and role of ICT in teaching are two prominent factors hindering teacher's readiness and confidence in using ICT support (Adewole, Akinwale & Omokanye, 2008; Aremu & Adediran, 2011). Problems such as poor ICT infrastructure are common challenges faced in the adoption of E-learning in Nigeria (Japheth & Cyprian, 2013). Nwana (2012) revealed that the few available E-learning platforms are not utilized because the teachers lack the knowledge and skills of computer application. This was buttressed by other studies that indicated that there is dearth of trained teachers for E-learning, lack of facilities, infrastructures and equipment's (Adu, Eze, Salako & Nyangechi, 2013; Ikemenjima, 2005; Jegede & Owolabi, 2008). Ajayi & Ekundayo (2009) revealed that some ICT challenges in secondary schools as: irregular power supply, inadequate computer literate teachers, high cost of purchasing computers in schools, inadequate facilities to support full application of the ICT and lack of fund.

Previous studies have been able to examine the availability and challenges of E-learning in Nigerian secondary schools as it have affected its adoption. However, these studies were largely theoretical studies and descriptive with limited empirical and inferential evidences. Furthermore, few previous studies as examined the public and private secondary schools vis-à-vis the E-learning facilities available to students in those school. It was these gaps that propelled the researcher to examine E-learning adoption among selected secondary schools in Ago-Iwoye, which comprises of both privately and publicly owned schools. Specifically, the study intends to expose difference between E-learning facilities available in Public and private secondary schools; whether there is a significant relationship between the availability of E-learning facilities and its application; and the relationship between the challenges of E-learning and its application.

THEORETICAL UNDERPINNING

Constructivism Learning Theory

This theory was propounded by Jerome Bruner (1960) whose work was influenced by Jean Piaget, a developmental psychologist. The underlying assumption of constructivism learning theory is the role which experiences with the atmosphere play in student knowledge acquisition. He argued that two key concepts within the constructivism learning theory which create the construction of an individual's new knowledge are accommodation and assimilation. Assimilating affords an individual to inculcate new experiences into the old one. This leads a student to develop new outlooks, rethink what were once misunderstood, and evaluate what is important, ultimately altering their perceptions (Bruner, 1960). Accommodation, on the other hand, is reframing the world and new experiences into the mental capacity already present. Individuals conceive a particular fashion in which the world operates (Bruner, 1960).

The role of teachers in the key assumptions/concepts of this theory i.e., accommodation and assimilation, is very important. The teachers according to this theory function as facilitators whose role is to aid the student when it comes to understanding what there are being taught. The resources and lesson plans that must be initiated for this learning theory take a very different approach toward traditional learning as well. Instead of been dependent on someone else's information and accepting it as truth, the constructivism learning theorists argues that students should be exposed to primary sources of knowledge acquisition so that they can learn from the incorporation of their own experiences. The constructivism learning theory will allow children to, at an early age or a late age, develop the skills and confidence to analyze the world around them, create solutions or support for developing issues, and then justify their words and actions (Bruner, 1960). This is done in such a way that it encourages those around them to do the same and respects the differences in opinions for the contributions that they can make to the whole of the situation. Classroom applications of constructivism support the philosophy of learning which build a students' understanding.

These reaffirmed the importance of application of E-learning facilities by teachers in students learning processes. The application of ICT'S as an essential tool will allow children to, at an early age; develop the skills and confidence to analyze the world around them since they are exposed to the use of technological resources. The utilization of ICT by students allows them to absorb and recall knowledge acquired from different technological sources. By so doing, students will be will equipped with the needed technological knowledge when they grow. Infact, employers may likely not to spend much money on ICT trainings. Creativity which is an essential tool for national development will be achieved.

METHODS AND MATERIALS

- **Study Area:** The study was carried out in selected secondary schools in Ijebu-North Local Government in Ogun state. This comprised of both 5 private and 5 government owned schools. These schools were considered because they are government approved, and as part of the ethical issues of the study, the names of the schools were not released.
- **Research Design**: This research study was descriptive in nature. A self-designed questionnaire which addresses the research objectives was used to collect the data for the study
- **Study Population and Sample Size:** The population for the study consisted of all the teachers and principals in selected government approved secondary schools in Ijebu- North, Ogun State. According to a pilot survey done by the researcher, there are 20 principals including their vices and 97 secondary school teachers. Therefore, the study population was one hundred and seven

(107) where 19% consist of the principals and 81% consist of the teachers. The sample size for the study was selected through the adoption of Yaro Yemen formula

$$n = \frac{N}{\left[1 + N\left(e\right)^2\right]}$$

where

n = Sample size,
N = Study population,
e = Margin of error.

$$\frac{107}{\left[1 + 107\left(0.05\right)^2\right]} = 84 \ respondents\left(Approximately\right)$$

The sample for the study was made up of 68 teachers (81%) and 16 principals (19%) who were randomly selected. The characteristics in terms of sex, age and size of the study population was put into consideration before the sample size was chosen. This was done to ensure the adequacy and representativeness of the sample size.

- **Instrument of Data Collection:** The study utilized both primary and secondary method of data collection. The primary data was sourced through the administration of a 34-item self-designed questionnaire. However, the secondary data was sourced from journals, conference presentations, and textbooks and so on.
- **Validity and Reliability of Research Instrument:** The questionnaire was assessed by selected experts of ICT studies. Comments and suggestions proffered by these experts were incorporated before the final draft of the instrument was made. In addition to this, a pilot study, consisting of 20% of the total sample size (i.e., 17 copies of questionnaire) was conducted to test the level of reliability of the items in the instrument, especially for the availability, application and challenges scales through the adoption of Cronbach's Alpha statistics. Results show the Cronbach Alpha of 0.80, 0.86 and 0.71 for availability, application and challenges scales. The benchmark, which was adopted, was that of George & Mallery (2003) provides the following rules of thumb: "_ >.9 – Excellent, _ >.8 – Good, _ >.7 – Acceptable, _ >.6 –

Questionable, _ >.5 – Poor and _ <.5 – Unacceptable" (p. 231). Therefore, all scales employed in this study were reliable.

- **Method of Data Analysis:** Data gathered from respondents were analysed through quantitative methods. Precisely, descriptive and inferential statistics were employed for the analysis of the data gathered. The descriptive statistics made use of frequency counts and simple percentages, while ANOVA and Regression analysis were used to pass inference on the gathered data. This was done using Statistical Package for Social Sciences (SPSS) 20.0 version.

DATA PRESENTATION AND ANALYSIS

Interpretation

Table 1 shows that respondents in private secondary schools were mostly surveyed (53.1%) followed by those in public schools (46.9%). 51.6% (33) of the respondents were male, while 48.4% (31) were female. Furthermore, the table shows that most that were surveyed were between the ages of 21 – 30. But in terms of religion, 48.4% (31) of the respondent were Christians, while 46.9% (30) were Muslims and about 4.7% (3) of the respondent practices other forms of religion.

Most of the respondents were married (59.4%), while 35.9% (23) of the respondents were single. In terms of ethnicity, 51.6% (33) of the respondents were Yoruba, 40.6% (26) were Igbo, 3.1% (2) were Hausa while 4.7% (3) fall into the category of other ethnic groups like Tapa, Isoko and Isobo. Meanwhile, the education section shows that 9.4% (6) of the respondent were O'Level holders, 59.4% (38) were OND/NCE/ Diploma holders, 23.4% (15) were B.Sc./HND holders, and 3.1% (2) were postgraduate degree holders while other (4.7%) were held certain professional certifications.

Table 1 also depicts that 40.6% (26) of the respondents have been in their profession between 1 – 5 years, while 43.8% (28) claimed 6 – 10 years and 15.6% (10) has been in the teaching profession for the period of 11 years or more. Lastly, 20.3% (20.3) earned 18,000 or less while half of the respondents (50.0%) earned 19,000 – 30,000 and 29.7% (19) earned more than 30,000.

Interpretation

Table 2 shows that 14.7% (5) of respondents in the private school claimed educational software are available in their school, while 3.3% (1) confirmed its availability in public schools. While 29.4% of the respondents had laptop in private school, only 3.3% (1) had in public. Furthermore, 38.2% (13), 26.5% (9), 20.6% (7), 17.6% (6), 8.8% (3) and 5.9% (2) of the respondent in private schools confirmed the availability

Table 1. Socio-demographic characteristics of respondents

Question	Response	Freq.	Percent
School Ownership Structure	Private	34	53.1
	Public	30	46.9
	Total	64	100.0
Sex	Male	33	51.6
	Female	31	48.4
	Total	64	100.0
Age	20 Years or Less	10	15.6
	21-30	32	50.0
	31-40	12	18.8
	Above 40 Years	10	15.6
	Total	64	100.0
Religion	Christianity	31	48.4
	Islam	30	46.9
	Others	3	4.70
	Total	64	100.0
Marital Status	Single	23	35.9
	Married	38	59.4
	Others	3	4.70
	Total	64	100.0
Ethnicity	Yoruba	33	51.6
	Igbo	26	40.6
	Hausa	2	3.10
	Others	3	4.70
	Total	64	100
Education	O'Level	6	9.40
	OND/NCE/DIPLOMA	38	59.4
	B.Sc./HND	15	23.4
	Post graduate	2	3.10
	Professional Qualification	3	4.70
	Total	64	100.0
Working Experience	1-5 years	26	40.6
	6-10 years	28	43.8
	11 years above	10	15.6
	Total	64	100.0
Monthly Income	18,000 or Less	13	20.3
	19,000-30,000	32	50.0
	More than 30,000	19	29.7
	Total	64	100.0

Source: Field Survey, 2015.

of E-learning facilities such as television, digital camera, projector, Internet facility, interactive boards and virtual library respectively. However, none these are available in the public counterparts. The table also shows that 23.5% (8), 32.4% (11) and 17.6% (6) of the respondent confirmed the availability of multimedia facilities, photocopiers and disc player respectively in private school while just 3.3% (1) of the respondent confirmed the aforementioned in public schools.

Table 2. Availability of e-learning facilities in secondary schools

Facilities	Private			Public		
	A	N.A	Total	A	N.A	Total
Educational Software	5 (14.7%)	29 (85.3%)	34 (100.0%)	1 (3.3%)	29 (96.7%)	30 (100.0%)
Laptop	10 (29.4%)	24 (70.6%)	34 (100.0%)	1 (3.3%)	29 (96.7%)	30 (100.0%)
Printer	15 (44.1%)	19 (55.9%)	34 (100.0%)	3 (10.0%	27 (90.0%)	30 (100.0%)
Television	13 (38.2%)	21 (61.8%)	34 (100.0%)	- (0.0%)	30 (100.0%)	30 (100.0%)
Digital Camera	9 (26.5%)	25 (73.5%)	34 (100.0%)	- (0.0%)	30 (100.0%)	30 (100.0%)
Projector	7 (20.6%)	27 (79.4%)	34 (100.0%)	- (0.0%)	30 (100.0%)	30 (100.0%)
Internet Facility	6 (17.6%)	28 (82.4%)	34 (100.0%)	- (0.0%)	30 (100.0%)	30 (100.0%)
Interactive Boards	3 (8.8%)	31 (91.2%)	34 (100.0%)	- (0.0%)	30 (100.0%)	30 (100.0%)
Virtual Library	2 (5.9%)	32 (94.1%)	34 (100.0%)	- (0.0%)	30 (100.0%)	30 (100.0%)
Multimedia Facilities	8 (23.5%)	26 (76.5%)	34 (100.0%)	1 (3.3%)	29 (96.7%)	30 (100.0%)
Photocopiers	11 (32.4%)	23 (67.6%)	34 (100.0%)	1 (3.3%)	29 (96.7%)	30 (100.0%)
Disc Player	6 (17.6%)	28 (82.4%)	34 (100.0%)	1 (3.3%)	29 (96.7%)	30 (100.0%)

Source: Field Survey, 2015.

Interpretation

In summarily, the table shows that few respondents adopt information and communication technologies in knowledge impartation in the surveyed secondary schools. Precisely, 40.6% (26) of the respondents do make use of instructional platforms in the teaching process while 59.4% (38) has never made use of the aforementioned ICT's in teaching. This becomes worrisome as Table 3 also shows that 71.9% (46), 65.6% (42) and 70.3% (45) of the respondent have never made use of projector, educational media and laptops/computer respectively in teaching. Furthermore, only about 40.6% (26) and 48.4% (31) of the respondents have made use of radio/television and pictures in imparting knowledge on students. However, the table shows that the larger percentage of the respondent (51.6%) does made use of online articles/platforms in teaching.

Table 3. Level of teachers' application of e-learning facilities

Questions	Freq.	Percent.
I use Instructional Platforms in my teaching process.		
Often	8	12.5
Sometimes	18	28.1
Never	38	59.4
Total	64	100.0
I make use of projector when teaching.		
Often	2	3.1
Sometimes	16	25.0
Never	46	71.9
Total	64	100.0
I do produce educational media for my own lesson process		
Often	2	3.1
Sometimes	20	31.3
Never	42	65.6
Total	64	100.0
I use Computer/Laptop to teach my students		
Often	3	4.7
Sometimes	16	25.0
Never	45	70.3
Total	64	100.0
I do use radio and television programmes to teach my student		
Often	2	3.1
Sometimes	24	37.5
Never	38	59.4
Total	64	100.0
I make use of pictures to impart knowledge on my students		
Often	6	9.3
Sometimes	25	39.1
Never	33	51.6
Total	64	100.0
I make use of online articles/platforms in teaching		
Often	2	3.2
Sometimes	31	48.4
Never	31	48.4
Total	64	100.0

Source: Field Survey, 2015.

Interpretation

Table 4 displays the challenges of E-learning adoption in Nigeria, specifically in selected Ogun state secondary schools. The table shows that 48.4% (31) of the respondent claimed that shortage of relevant books has posed several challenges

Table 4. Challenges of e-learning adoption

Questions	Freq.	Percent
Shortage of relevant books have pose several challenges on the adoption of information and communication technologies in schools		
Agree	31	48.4
Disagree	26	40.7
Undecided	7	10.9
Total	64	100.0
Epileptic power supply has hindered the use of information and communication technologies in schools		
Agree	46	71.9
Disagree	12	18.9
Undecided	6	9.4
Total	64	100.0
Teachers' resistance to change has affected the use of information and communication technologies in schools		
Agree	25	39.1
Disagree	14	21.8
Undecided	25	39.1
Total	64	100.0
Inadequate funding is a challenge to information and communication technologies in secondary schools		
Agree	36	56.3
Disagree	13	20.3
Undecided	15	23.4
Total	64	100.0
Shortage of skilled personnel's in information and communication technologies has hindered its usage in secondary schools		
Agree	31	48.4
Disagree	19	29.7
Undecided	14	21.9
Total	64	100.0

Source: Field Survey, 2015.

on the adoption of ICT enabled teaching and learning process. These ranges from computer manuals on how to operate a to that of the educational software's. 71.9% (46) argued that epileptic power supply has hindered the use of E-learning facilities. Though teacher's resistance to change might not be considered as a challenge for E-learning adoption, however, challenges such as funding (56.3%) and shortage of skill personnel (48.4%) has hindered the adoption of E-learning in secondary schools.

H_1: There will be significant difference between E-learning facilities available in private and public secondary schools

Table 5. Test of hypotheses

Independent Samples Test										
		Levene's Test for Equallty of Variances		t-Test for Equality of Means						
									95% Confidence Interval of the Difference	
		F	Sig.	t	df	Sig. (2-Tailed)	Mean Difference	Std. Error Difference	Lower	Upper
E-learning Facilities	Equal variances assumed	50.484	.000	4.461	62	.000	2.586	.580	1.427	3.745
	Equal variances not assumed			4.717	37.785	.000	2.586	.548	1.476	3.696

Source: Field Survey, 2015.

Table 6.

Coefficients[a]						
Model		Unstandardized Coefficients		Standardized Coefficients	t	Sig.
		B	Std. Error	Beta		
1	(Constant)	6.640	.721		9.207	.000
	E-learning Facilities	**.771**	.233	.387	3.304	**.002**

Source: Field Survey, 2015.

a. Dependent Variable: Teachers' Application.

Interpretation

Table 5 shows that the significance value for the tested hypothesis is 0.00 which is less than 0.05 (i.e. $p<0.05$). This means that there is a significant difference between E-learning facilities available in private and public secondary schools. Therefore, the hypothesis is accepted.

H_2: There will be a significant relationship between availability of E-learning facilities and teachers application.

Table 7.

Coefficients[a]						
Model		Unstandardized Coefficients		Standardized Coefficients	t	Sig.
		B	Std. Error	Beta		
1	(Constant)	13.289	2.186		6.078	.000
	challenges	**-1.001**	.266	-.431	-3.763	**.000**

Source: Field Survey, 2015.
a. Dependent Variable: Teachers Application

Interpretation

Table 6 shows the regression analysis for the tested hypothesis. The significance value is 0.002 which is less than 0.05 level of confidence i.e., P<0.05. Therefore, there is a positive significant relationship between availability of E-learning facilities and teachers application. By extension, as these aforementioned facilities are readily made available, there will be a corresponding increase in teacher's application to teaching and learning process. Hence, the hypothesis is accepted.

H_3: There will be significant relationship between challenges of E-learning adoption and its application

Interpretation

The significance level for the tested hypothesis as displayed in the table below is 0.000 which is less than 0.05 (i.e., P<0.05). This connotes that there is a negative significant relationship between Challenges of E-learning adoption and teacher's application. In essence, the more the challenges of E-learning adoption prevail, the lesser teacher's application of E-learning facilities application to teaching and learning. Therefore, the hypothesis is accepted.

DISCUSSION OF FINDINGS

The study examined E-learning in Ogun state, specifically secondary schools in Ijebu-North Local Government, which comprised of both privately and publicly owned schools. The study found out that E-learning facilities such as educational software, laptop, printer, Television, digital, camera, projector, Internet facility, interactive boards, virtual library, multimedia facilities, photocopiers and disc player are not readily available in the surveyed secondary schools especially in the government owned secondary schools. This corroborates the findings of Global Information Technology Report (2005), Effiong, (2005); Jegede & Owolabi (2008); Seiden (2000); Uhaegbu, (2001) and Abodcrin & Kumuyi (2013) who submitted that there is shortage of E-learning tools in most schools found in Nigeria.

Furthermore, the study found out that there is a significant difference between E-learning facilities available in private and public secondary schools. By implication, this might be linked with the reason private school students are more intelligent and compete successfully than their counterparts in public schools since they are more exposed to the use of information and telecommunication technologies. Hence, this corroborates the assumptions of *Constructivism Learning Theory* which opined that experiences with the adjoining atmosphere play in student education. Precisely, the constructivism learning theory supports that students should be exposed to data, primary sources, and the ability to interact with other students so that they can learn from the incorporation of their experiences.

This study also found out that only few teachers apply the available E-learning facilities in teaching process, except for electronic articles that most of the resort to in lieu of textbooks. Furthermore, there was a positive significant relationship between availability of E-learning facilities and teachers application. Therefore, since there is a dearth of these facilities in the surveyed schools, it is right to say that teacher's application to teaching will be minimal.

Finally, the study found out that shortage of relevant books has posed several challenges on the adoption of ICT enabled teaching and learning process. This includes computer manuals amongst others. Furthermore, challenges such as epileptic power supply shortage of skill personnel have hindered the adoption of E-learning in secondary schools. This was similar to the findings of Adewole, Akinwale & Omokanye (2008), Aremu & Adediran (2011), Japheth & Cyprian (2013), Ikemenjima (2005) Jegede & Owolabi (2008) Adu, Eze, Salako & Nyangechi (2013) and Ajayi & Ekundayo (2009) who argued that there is dearth of trained teachers for E-learning, lack of facilities, infrastructures and equipment in Nigerian schools. More so, there was a negative significant relationship between challenges of E-learning adoption and teachers' application. Hence, the more the challenges faced by teachers as regards E-learning, there will be a corresponding decline in their application to teaching and learning process.

CONCLUSION AND RECOMMENDATIONS

The study concludes that despite the roles ICT-driven instructional aids can play in knowledge acquisition, both at formal and informal levels, Nigerian schools, especially the government owned, are yet to extensively avail and adopt them for learning process as a result of certain challenges. Owning to this, the following recommendations were made to government, school management and all other concerned bodies

1. More investment is to be made to education especially in the aspect of information and telecommunication technologies. This will ensure adequate availability of E-learning facilities for teachers and student use alike.
2. A massive computer literacy training program is to be embarked upon nation-wide particularly for teachers and learners at secondary school level. This should be accomplished by training teachers through workshops, seminars, and conferences. This will enhance teacher's application of E-learning to teaching process.
3. Teachers in Nigeria should be motivated and encouraged to develop interest for the use of multimedia devices and software relevant to teaching and learning process.
4. Technologists and technicians should be employed to take care of these E-learning facilities and carry out maintenance routines so as to ensure sustainable ICT adoption within the context of secondary schools.
5. Standby generators, Uninterruptable Power Supplies (UPS) devices and inverters are to be made available to tackle the problem of unstable power supply. This will encourage use of e-learning platforms in secondary schools

REFERENCES

Abimbade, A. (2002). Perspective of Technology Integration and Effectiveness of Computer Assisted Instruction (CAI) in Primary Mathematics Classroom. *Uniqua Research Chronicle*, *4*(2), 88–107.

Aboderin, O., & Kumuyi, G. (2013). The Problems and Prospects of E-learning in Curriculum Implementation in Secondary Schools in Ondo State, Nigeria. *International Journal of Educational Research and Technology*, *4*(1), 90–96.

Adewole, A., Akinwale, A., & Omokanye, B. (2008). A Consulting Information and Communication Technologies (ICTs) Teacher Model for Teaching Secondary School Curriculum.*Proceedings of the third conference on science and National Development* (vol. 38, pp. 23-29). IbadanUniversity Press

Adu, E., Eze, I., Salako, E., & Nyangechi, J. (2013). E-learning and Distance Education in Nigeria. *International Journal of Science and Technology*, *2*(2), 203–210.

Ajayi, G. (2005). Information and Communication Technology: Building capacity in African Universities.*Conference proceedings*. University of Nairobi.

Ajayi, I., & Ekundayo, H. (2009). The application of information and communication technology in Nigerian secondary schools. *International NGO Journal*, *4*(5), 281–286.

Apple Computer. (2002). *The Impact of Technology on Students' Achievement*. Retrieved on 10th February, 2016 from http://www.apple.com/education/research/index2

Aremu, A., & Adediran, E. (2011). Teacher Readiness to Integrate Information Technology into Teaching and Learning Processes in Nigerian Secondary Schools: A Case Study. *African Research Review*, *5*(4), 178–190. doi:10.4314/afrrev.v5i4.69275

Arinze, F., Okonkwo, E., & Iwunor, A. (2012). ICT Application in Secondary Schools and Students Academic Performance in Social Studies. *African Research Review*, *6*(4), 266–278. doi:10.4314/afrrev.v6i4.18

DfES. (2003). *Towards a Unified E-learning Strategy, London*. Retrieved on 24th April, 2016 from http://www.fmegovng.org/

Effiong, J. (2005). Business education in the era of information and communication technology: Issues, problems and prospects. *Business Education Journal*, *5*(1), 3–7.

George, D., & Mallery, P. (2003). *SPSS for Windows step by step: A simple guide and reference*. Boston: Allyn & Bacon Publishing.

Ikemenjima, D. (2005). *E-Education in Nigeria: Challenges and prospects*. A Paper presented at the 8th UN ICT Task Force Meeting, Dublin, Ireland.

Japheth, T., & Cyprian, C. (2013). The Impact of ICT-Driven Instructional Aids in Nigerian Secondary Schools. *International Journal of Basic and Applied Sciences*, *1*(3), 511–518.

Jegede, P., & Owolabi, A. (2003). Computer education in Nigerian secondary schools: Gaps between policy and practice. *Meridian: A Middle School Technology Journal*, *6*(2), 1 – 11

Nwana, S. (2012). Challenges in the application of E-learning by secondary school teachers in Anambra State, Nigeria Africa. *Journal of Teacher Education*, *1*(1), 32–41.

Rozina, I. (2002). Theory and Practice of Computer-assisted Communication in Russia: Present-day Situation and Future Perspectives. In: Theory of Communication and Applied Communication. *Journal of Russian Communication Association*, *1*(4), 1–12.

Seiden, P. (2000). Where have all the patrons of ICT gone. *Reference and User Services Quarterly*, *39*(3), 2–10.

Uhaegbu, A. (2001). *The information user: Issues and themes*. Enugu: John-Jacobs Publishers.

ADDITIONAL READING

Evey, C., Emmanuel, O., Joseph, A., Denis, U., & Asinde, F. (2010). The need for ICT in secondary schools curriculum in Nigeria. *Nigerian Journal of Teacher Education and Teaching*, *8*(2), 183–192.

Inyang, A. M. (2004). Essentials of Educational Technology. Calabar: Wusen Publishers Limited, Nigeria.

Kozma & Anderson. (2002). Qualitative Case Studies of Innovative Pedagogical Practices Using ICT. *Journal of Computer Assisted Learning*, *18*(3), 387–394.

KEY TERMS AND DEFINITION

Adoption: The use of electronic devices in teaching and learning process.

Challenges: Obstacles faced by teachers in adopting electronic platforms in teaching and learning process.

Classroom: A learning space comprising of teachers, students and other relevant materials for the purpose of knowledge acquisition.

Education: The acquisition of knowledge, skills, values, beliefs, and habits, which might be formal or informal in nature.

Electronic Platform: A channel through which students are been taught through the use of electronic devices.

Information and Communication Technology (ICT): An umbrella term used to describe any electronic devices or application.

Knowledge Acquisition: The process of absorbing and storing new information in memory through electronic means.

Chapter 6
Potentials of Digital Assistive Technology and Special Education in Kenya

Foad Hamidi
York University, Canada

Michaela Hynie
York University, Canada

Patrick Mbullo Owuor
Maseno University, Kenya

Melanie Baljko
York University, Canada

Susan McGrath
York University, Canada

ABSTRACT

Technology specifically designed for people with disabilities is important in lowering boundaries to education, employment and basic life needs. However, the growth of a vibrant tech sector in Kenya has had little effect on the prevalence of digital assistive technology in the country. In this chapter, the authors report on initial explorations undertaken in Kisumu, Kenya to identify existing strengths, relationships, and gaps in access to digital assistive technology. The goal was to explore opportunities for initiatives in participatory design of assistive technology, using an international community/academic partnership. Relevant literature and projects from the areas of Information and Computer Technology for Development (ICT4D), Human-Computer Interaction (HCI) and Critical Disability Studies are reviewed and, these theories are grounded in the authors' experience working with stakeholders in the region. The conclusion discusses promising future directions for participatory and collaborative research in Kenya, and more broadly in the East African context.

DOI: 10.4018/978-1-5225-2565-3.ch006

INTRODUCTION

The availability of non-commercial assistive technology is improving. A series of factors is driving this change including the upsurge in commons-based production (including open-source software and hardware), the growth of Do-It-Yourself (DIY) and Maker communities, and a proliferation of accessible and affordable prototyping components (Hurst & Kane, 2013). This improved access is especially promising in contexts where the need has long been recognized but technological and economical barriers have limited the development and deployment of Assistive Technology (AT) relevant to localized contexts and communities of users.

In this chapter, we will discuss the potential promising directions of digital AT development in Kenya, with a view to the larger East African context. We review relevant literature and projects from the fields of Information and Communication Technology for Development (ICT4D), Human-Computer Interaction (HCI), Critical Disability Studies and Participatory Design (PD) and related theories to researchers experiences working with stakeholders in the region. We conclude with a discussion of promising future directions. This chapter emerges out of a collaborative project being undertaken by an international, interdisciplinary team of social scientists, computer scientists and community agencies working together to develop a transnational model to improve access to AT. The transnational team includes two computer scientists specializing in digital AT and ICT4D, a psychologist and mental health researcher, an expert in social work and refugee studies and a founding member of a Kenyan NGO with experience working with children with disabilities. Our team shares a strong interest in interdisciplinary collaboration and community engagement with previous experiences working in East Africa and other developing regions of the world including South Asia and Central America.

This work investigates the use of cross-sectorial academic/community collaborations to improve the well-being of children with disabilities in Kenya, and ultimately, for all who might benefit, both in Kenya and around the world. Academic/community collaborations have tremendous potential in addressing complex social issues (Boyer, 1990; Chernikova, 2011). This is particularly the case if community members who are directly affected by the issues being studied are engaged (Minkler, 2005). These ideas of participatory and collaborative work underpin and shape the current initiative.

As part of this project, a stakeholder meeting, organized and facilitated by local community partner was held in the summer of 2014 in Kisumu. The stakeholders included representatives from different government, non-government, and community agencies and a national advocacy group. The meeting had three goals. First, to create a network map to explore existing, potential and desired relationships between different stakeholders; second, to identify gaps and barriers in accessibility and disability at the policy and the intervention levels; and third, to discuss future plans and directions for the collaboration. Several potential directions for development were identified. These included: public-private partnerships, routes of e-citizenship for people with disabilities, creation of employment opportunities for people with disabilities, and improved resources for people with mental health and cognitive disabilities. Additionally, several visits were made to schools for children with disabilities in the Kisumu region of Kenya to observe their computer facilities. The goal of this step was to gain preliminary first-hand knowledge about the current use of digital assistive technology in a variety of schools. Focus was on digital AT, which involves the use of computer technology, as opposed to the broader definition of AT as covering anything that might be used to compensate for lack of certain abilities (Reed & Bowser, 2005; WHO, 2009).

THE POTENTIALS OF DIGITAL ASSISTIVE TECHNOLOGY IN KENYA

In recent years, Kenya has gained worldwide recognition as a leader in the field of Information and Communication Technology (ICT) in East Africa (Smith, 2012). Through the global success and adoption of homegrown technologies such as the Ushahidi (http://www.ushahidi.com, an open-source crowdsourcing information mapping platform) and m-Pesa (https://www.mpesa.in, a mobile phone banking and money transfer system). Kenya has emerged as a leader not only in adoption and use of technology but also in its innovation and development. Additionally, startup business incubation and training centers such as the iHub (www.ihub.co.ke) and m:lab (mlab.co.ke) have provided a launching pad for many app developers and Kenya has been steadily producing online apps targeting various user populations ranging from farmers, using M-Farm (www.mfarm.co.ke/),to fine diners, using Eat Out (www.eatout.co.ke). Finally, several Kenyan software companies that focus on financial services and digital payment systems such as Craft Silicon (http://www.craftsilicon.com) and Cellulant (www.cellulant.com) have gained great success with contracts in other countries.

This atmosphere has prompted some to call Kenya the "Silicon Savannah" (Smith, 2012); a notion that has its critics point out that the focus on Kenya's ICT successes has created a reliance on a "grant economy" rather than an "investment economy" with interest in short-term gains rather than long-term projects (Mulupi, 2013). Comparing Kenya with neighboring Rwanda, a possible rival with its own program to strengthen its ICT sector through infrastructure implementation and educational programs, some critics have pointed out that Kenya needs to provide more foundational and educational support for its new generation of technology entrepreneurs (Bezy, 2013). In particular, while Carnegie Mellon University (CMU) has recently initiated a professional certification program for software developers in Kenya (Mbuvi, 2012), it established a more in-depth on-site engineering Master's program in Kigali, Rwanda with CMU faculty in residence (Bezy, 2011). Similar programs with focus on ICT training would strengthen Kenya's position in the future. Since 2013, the Kenyan government has put forward an ambitious initiative, the Digital Learning Programme (DLP) that aims to provide laptops for children in the country's primary schools (http://digischool.icta.go.ke/). The program has proven controversial with some observers drawing attention to difficulties in its effective deployment (e.g., Matinde, 2015). Despite the criticisms, the variety and vitality of Kenya's ICT initiatives, projects and businesses point to a promising future; a future that has potential to bring about better conditions for some of the nation's most vulnerable citizens. This positive attitude towards technology use and taking of steps towards providing greater access to computers and the Internet for the general population in Kenya, Rwanda, and to a lesser degree Uganda and Tanzania, have prompted research initiatives to identify further possibilities for supporting education and governance initiatives through technology in the countries in this region (Sike et al., 2014; Farell, 2007; IST-Africa, 2016).

The field of information and communication technologies for development involves the design, development and deployment of digital technologies in the context of developing countries and with a view to support empowerment and inclusion in addition to financial and social development (Unwin, 2009). According to the UNESCO, ICTs can refer to "technologies that are used to transmit, store, display, share or exchange information by electronic means" (UNESCO, 2007). As a field, ICT4D has matured considerably over the last few decades (Chepken et al., 2012). This is not only apparent in the proliferation of leading Information Technology (IT) companies in developing countries, as illustrated through the Kenyan examples above, but also in the emergence of progressive and socially-informed human computer interaction design methodologies and approaches (Marsden, 2008). New design approaches take into consideration prevalent technologies and practices in a given context and build systems on top of existing infrastructure (e.g., mobile cell phones), generating projects that are informed by the cultural and social context in which they are to be used (Marsden, 2008). Examples of such designs have become more common in recent years and include the innovative use of affordable cell phones to aid with information sharing and education (Marsden, 2008), as well as e-citizenship (Dodson, 2013). These examples are particularly relevant as recent research shows a high adaption and popularity of affordable cell phone use in Kenya for resource sharing and family coordination (Oduor et al., 2014).

Additionally, Participatory Design and Co-design approaches have gradually gained more and more recognition for their capacity to take into account users' domain knowledge and inform design through engaging and collaborating closely with stakeholders (Sanders & Stappers, 2008). This shift has had important implications for ICT4D projects as they recognize the importance of incorporating multiple points of view in design and being sensitive to different, sometimes, conflicting values. The challenge of intercultural collaboration in these contexts means that establishing trust and mutually beneficial relationships with members of the user population (i.e., *human access points*) is deemed critical to project success (Marsden et al., 2008). In developing countries, NGOs provide many valuable resources and much insight. Some researchers have argued that these grassroots organizations should be involved in all phases of projects from initial assessment to implementation, to evaluation and follow-up (Gitau & Marsden, 2009). Finally, new methods encourage designers to use working prototypes to move projects from the lab into the field early on in order to gain relevant hands-on knowledge in the domain (Rogers, 2011). The insights and experiences of the ICT4D community are rooted in many decades of experience and can provide valuable directions and motivation for digital assistive technology development in the future.

Participatory Design and Assistive Technology

Participatory Design (PD) is a design methodology that emphasizes the incorporation of user domain knowledge and recognizes the importance of collaborating and co-creating with the user and their community (Greenbaum & Kyang, 1991; Schuler & Namioka, 1993). An essential technique of PD is to actively engage the user of the technology in its design. Originally, PD was used to develop technologies for the workplace; however, it became apparent that it also offers great value to many other areas of design and development including designing with users with disabilities (Kensing & Blomberg, 1998) and specifically designing with and for children with disabilities (Hourcade et al., 2014; Frauenberger et al., 2012).

Participatory Design is an approach within Community Based Research (CBR), a series of approaches and methods that originate from research in a community's needs and interests and aim to bring about social change as part of the research process (Horowitz et al., 2009). CBR incorporates a range of methodologies including Action Research, which focuses on collaborative research between participants and researchers to collectively solve social problems, and Participatory Action Research, that explicitly includes critical theory stances and awareness of larger social forces into the collaborative research process. What these approaches all share is a commitment to grounding the work in the needs of the community, shared decision making between community members and researchers, engagement of community members in multiple aspects of the research process, and creating concrete changes for the community both through the outcomes of the research and through participation in the process itself (Horowitz et al., 2009; Israel et al., 1998; Kassam & Tetty, 2003; Minkler, 2005; Suarez-Balcazar et al., 2005). CBR is a relevant approach in the context of ICT4D as it is explicitly democratic, collaborative and interdisciplinary. Furthermore, it is cyclical and aims to implement action iteratively, incorporating stakeholder feedback and insights at each step. In recent works, the potential of CBR for research in human computer interaction is recognized (Hayes, 2011) while previous research projects recognize the relevance of PD for the development of AT. (Frauenberger et al., 2012; Hourcade et al., 2014). Many designers of assistive technology do not have direct experience of the physical and/or mental conditions that necessitate their designs. This may be the case even for those researchers who have disabilities themselves and are themselves users of assistive technology. Each person's use scenario may be unique and it is important to develop and examine methods that allow for the incorporation of different perspectives and experiences into the tailored designs.

This necessity for tailored designs calls for the development and application of design methods that allow for customization and direct involvement of representatives of user population in design. It is essential for technology designers to work closely with representatives from the population they are designing for, so that they can have a better understanding of the potentials and perils of their creation when used by real people. Additionally, the importance of accommodating designers and co-designers who themselves have disabilities, and thus "insider knowledge", in the design of assistive technology is becoming increasingly apparent. PD can support the popular approach of *Universal Design* (or *Inclusive Design*) that aims to produce products and build environments that are accessible to the largest number of people (Preiser, 2001) by incorporating insights from people with disabilities into the design of alternative or customizable system configurations. Universal Design (UD) is known to cause a "curb-cut" phenomenon, which involves benefits to the larger society as a result of design decisions aimed originally at people with disabilities (Jacobs, 1999). The term refers to ramp cut into the curbs of sidewalks, originally designed to allow access to people who use wheelchairs, but which subsequently became useful to a wider group of people without disabilities, for example people pushing shopping carts or baby strollers, among others. Treviranus (2014) has argued that this recurring phenomenon provides additional rationale to support accessible initiatives that can benefit the larger community as a whole and provide more access to economic, as well as, social resources.

Many previous projects have used PD for the development of assistive technologies and have engaged adults with disabilities. Several projects have explored using PD with older adults with amnesia (Wu et al., 2004; Moffatt et al., 2004) and aphasia (Galliers et al., 2012; Wilson et al., 2015). Davidson and Jensen (2013) have shown that involving users in the design of AT can reveal new insights previously unknown to researchers. The researchers involved potential users in the design of a smartphone app to monitor health. Participants were divided into five groups, each of which worked on one of five possible applications. Analysis of the designs showed that the participants identified several novel health metrics (i.e., metrics that were not already included in existing apps). Also, the breadth of design suggestions indicated that the participants had a more holistic view of health that allowed them unique insights into the design. This holistic view was not completely in convergence with that of the technology designers (Davidson & Jensen, 2013). Wilson et al. (2015) argue that the development and use of "tangible design languages" (non-verbal design representations used to communicate ideas and feedback to researchers during the design process) is a crucial aspect of using this approach. They demonstrate their method by successfully engaging adults with aphasia in the design of two therapeutic systems, GeST and EVA Park. They identified images, photo diaries and scenarios, tangible avatars and high-fidelity prototypes as elements that can help create a common language between researchers and users with communication disorders (Wilson et al., 2015). Although most of these studies were conducted using adults with disabilities, rather than children with disabilities, the methodology is still relevant.

Recently, more participatory design projects are being conducted with children and specifically children with disabilities. The topic of designing for and with children with disabilities has been the topic of several international workshops in recent years (e.g., Hourcade et al., 2014). Frauenberger et al. (2012) have argued strongly for engaging children with disabilities and their community in the design and evaluation of computational interfaces, arguing that the three identified benefits of using Participatory Design, namely 1) better understanding of requirements, 2) building realistic expectations in target groups and 3) empowerment of marginalized groups, are amplified in this context.

Different approaches for the design of AT can be used to involve stakeholders such as children with disabilities and the adults who are familiar with them. Some approaches involve children as *informants*, whereby they provide ideas for the designs and help with evaluation but are not involved as full design partners. For example, Brederode et al. (2005) involved children with different physical and learning disabilities as informants in the development and evaluation of *pOwerball*, an augmented reality tabletop tangible game to encourage collaboration and social interaction for children with mixed abilities. Another approach, *PD with Proxies*, involves adults who work with the target population (e.g., speech language pathologists, special education teachers) or the children's caregivers or parents in the design process. For example, Kientz et al. (2007) used the help of adult proxies (caregivers, parents and teachers of children with autism) to inform the design of wearable interface to detect self-stimulatory behavior in children with autism. Finally, other approaches have included children with disabilities themselves directly in the design process. Examples include the ECHOES project (Frauenberger et al., 2012; Frauenberger et al., 2011), *ReacTikles* (Keay-Bright, 2007) and SIDES (Piper et al., 2006), in which researchers collaborated with children with varying degrees of autism to develop new systems.

Several research projects have been concerned with the development of new methods and tools for engaging children with disabilities in PD. Techniques include the use of technology probes (Dawe, 2007), templates, personas and computer mockups (Benton et al., 2011; Millen et al., 2011). In all of these approaches, researchers argue for the importance of including children with different abilities in the design process. In analyzing children's feedback after PD sessions, Anthony et al. (2012) found that the children felt empowered by being included in the design process and motivated to engage in discussions and hands-on activity.

The above examples, regardless of the diverse populations that have participated in PD for the development of assistive technology, illustrate two themes: first, the inclusion of users in the design process allows for the uncovering of valuable domain expertise (Visser et al., 2005) and, second, the challenge for designers of technology to come up with appropriate methods that allow for the expression and integration of this expertise is on-going (e.g., alternatives to spoken or written language, lo-fi and/or high-fi prototypes, etc.) (Sanders & Stappers, 2008). These results are consistent with recent developments in the theory of disability that call for approaches that support the increased inclusion and participation of people with disabilities.

Over the last decades, there has been a major shift in understanding disability and its social and personal implications (Rosenbaum & Stewart, 2004). This shift is readily apparent in the work of the World Health Organization (WHO), which has challenged the traditional *medical model of disability* and has developed alternative frameworks for characterizing the lived experience of disability (World Health Organization, 2001). In recent decades, disability activists and researchers have rejected the medical model and its Positivist roots, a model which emphasizes the negative consequences of disease as defining factors in disability and as one's own impairment as the locus of disability. As well, they have argued for the inclusion of the voice of people with disabilities at every stage of research and policymaking (Oliver, 1998).

There is a rich discourse about the experience of disability and alternatives to the medical model. Even early versions of the International Classification of Functioning, Disability and Health (ICF), developed and maintained by the WHO, took care to distinguish between, for instance, the notion of *disorder*, which pertains to physiological bodily function, and the notion of *participation restriction*, which pertains to an individual's experiences and ability in life situations, understood to arise from the dynamic interaction between the individual's own embodiment, personal factors, and environmental factors (World Health Organization, 2001). The medical model of disability places the locus of disability in an individual's own embodiment, whereas other models hold that the experience of disability is emergent in the interaction of the individual with the individual's context.

The experience of disability is emergent in the interaction of an individual in his or her context: this has become a core idea in Disability Studies. It can be seen in the *social model of disability* (Wendell & Linton, 2000). In this model, disability is not viewed as a condition attributed to an individual due to disease or impairment. Rather, it is viewed as a social construct representing the dynamic interaction of the individual and society (Rosenbaum & Stewart, 2004). The social model of disability recognizes contextual factors, such as the availability and quality of intervention programs, as important factors in understanding and addressing disability (Wendell & Linton, 2000). Other models that aim to both question and complement the social model by placing the subjective experience of disability front and center in debates on this topic are also emerging (Oliver, 1998; Frauenberger et al., 2010). A prominent approach is characterized as the postmodern model of disability (Corker & Shakespeare, 2002). Influenced by post-structuralism, this approach does not aim to replace the social model of disability but to supplement and enrich it by including "analyses of the subjectivity of those positioned by social structures and practices as having disabilities" (Corker & Shakespeare, 2002, p. xiv). The postmodernist approach aims to challenge the notion of a rational knowing subject and a universal mind and instead aims to recognize multiple subjects and multiple knowledges. Additionally, it argues that the modernist "focus on the individual as an autonomous agent needs to be deconstructed, contested, and troubled" (Corker & Shakespeare, 2002, p. 3). More concretely, the postmodern model draws parallels between notions of disability and other sources of social differences, such as sex and race, and argues for an inclusion of personal subjective experiences as well as socially constructed experiences (Galis, 2011). While this model is relatively new and its influence in the field of disability and specifically assistive technology is still to emerge, there are interesting parallels between its stance and a recent and influential paradigm in HCI research, known as Third Wave HCI (Harrison et al., 2007).

During its history, HCI has been approached from many angles, with perspectives informed from different disciplines such as Computer Science, Psychology, Engineering and Design. In recent years, researchers have identified three different paradigms in HCI, emerging in sequence one after another; these are not necessarily contradicting each other, but rather overlapping and complementing one another (Harrison et al., 2007). Harrison et al. (2007) categorize each paradigm based on the central metaphors that inform them and argue that the approaches, methods and points of focus that each promotes follows from that central metaphor. The first paradigm called 'Human Factors', views interaction as man-machine coupling and focuses on the goal of optimizing the fit between humans and machines. The second paradigm called ' Classical Cognitivist/Information Processing Paradigm', uses the idea of interaction as information communication as its central metaphor and aims to optimize the accuracy and efficiency of information transfer. The Third Paradigm views interaction as phenomenologically situated and focuses on support for meaning-making in specific contexts. With its focus on meaning creation and human experience expressed through multiple perspectives and the relationships among these perspectives, the Third Paradigm adds much-needed sensitivity called for by the social and post-modern models of disability.

Different models of disability have had profound effects on assistive technology research and researchers have argued for the raising of awareness around these models for assistive technology developers (Mankoff et al., 2010). As early as 1979, researchers called for technologies that avoided trying to "fix" people and making them "normal", but strived instead to recognize and incorporate their subjective experiences, interests and talents (Goldenberg, 1979). Over the years, this focus has continued and recently has matured into research approaches that recognize the existence of a "culture of disability" that should be respected, understood and incorporated by researchers (Benton et al., 2014).

Past: Digital Assistive Technology and Successful Design Approaches in Kenya

According to the Kenya National Survey for Persons with Disability (KNSPWD), 32% of people with disability use an assistive device of which 20% are information devices, 12% are personal mobility devices, 0.3% are communication aids, 0.1% are household items, 0.4% are personal and care protection and 0.1% are for products and goods handling (NCPD, 2008). The distribution of assistive devices is greater in the urban setting (42%) than in the rural (16%). This is attributed to the fact that information accessibility and availability of infrastructure is more established in urban settings. Nairobi has the highest (42%) use of assistive devices or support services compared to other provinces, while North Eastern Province has the lowest (16%). People with disabilities between the ages of 15–24 reported low use of assistive devices (15%). Low use of assistive devices was also recorded among people with disabilities with nursery or kindergarten level of education (9%).

Currently, most digital assistive technology initiatives have focused on software applications such as screen readers or virtual keyboards. Many of these solutions are proprietary and expensive. As a result, only a small number of children with disabilities get access to technology. The problem with this approach is that it leads to over-dependence on software that is unaffordable. Such an approach is just as ineffective whether the software runs on PCs, netbooks, or cell phones since even the best-known and most prevalent assistive technologies designed for these devices are extremely expensive.

Investment in open-source software that can be easily modified, shared and used in different contexts has been recommended in Kenya (Botelho, 2009). Botelho suggested that with recent infrastructural enhancement in Kenya and the lowering cost of hardware everywhere, the real obstacle is not hardware costs but a lack of awareness around the benefits of supporting and using Free Open-Source Software (FOSS) assistive technology solutions. Thus, companies, governments, NGOs, foundations, or individuals interested in using or contributing to assistive technology that is truly decentralized, scalable, affordable, and resilient should focus on finding and possibly developing free and open source software, open communication protocols, and open file and other standards, rather than lowering the cost of hardware. Additionally, public-private partnerships means the public sector supports private companies not only with financial support but with policy support that protects the rights of people with disabilities through the promotion and procurement of technology that is customizable and interoperable such that it can be incorporated into assistive technology solutions (Botelho, 2009).

Kenya is emerging as a leader not only in the ICT sector but also in social programs that incorporate technology. For example, in the Kilimani Primary School in Nairobi, digital assistive technology has been used to help students with visual impairment (Cullen, 2006). This school that has been providing instruction to children with visual impairment since the 1980s, recently started to replace its traditional Braillers (devices that can be used to write documents in Braille) with computer systems. Braillers, while useful, were expensive and easy to break. Additionally, Braille books were expensive and took a lot of space. Recently, in collaboration with Sight Savers International (www.sightsavers.org/, an organization strongly advocating inclusive education) and Computer Aid International (www.computeraid.org/, a non-profit organization that recycles old computers), Kilimani School introduced several digital assistive technology devices for their students. Initially, a refurbished computer with the proprietary Job Access With Speech (JAWS) program was used. JAWS is prohibitively expensive for the school and only a partially functional demo version was used. Observing this shortcoming, Sight Savers has been working on an open-source platform called the Dolphin Project, in which a suit of software designed to help users with visual impairment is provided on a USB stick. The suite includes magnification, voice and speech output and Braille support. The software is to be provided to both students at Kilimani School and Kenyatta University in Nairobi (Cullen, 2006).

Additionally, a scheme was developed to provide children who are marginalized by living in slums and low-income neighborhoods with computer access through shared computers that could be used via USB keys loaded with all the software needed to operate a system. The low-costing USB keys are given to children as storage for their personal systems and saved data and can be used at communal computer centers in slums. By creating a way to share rather than own expensive resources, this program aims to provide more affordable access to information and education resources for users (Simmons, 2014). With the prevalence of such initiatives, more children with disabilities will have access to more affordable technology in the near future. The challenge then is whether in addition to the new wave of affordable computers and laptops, they will also have affordable assistive technologies such as customized keyboards, switches and bridging software that allow them to use these new technologies.

An area of research with much potential for future exploration is designing assistive technology for users with mental health issues and cognitive disabilities. Designing for these populations poses special challenges since traditionally assistive technology has been more effective in helping users with physical disabilities. Many new directions are being explored in this area including therapeutic systems for adults with aphasia (Moffatt et al., 2004), safety systems for older adults with dementia (Gowans et al., 2004) and therapeutic and learning systems for children on the autism spectrum (Porayska-Pomsta et al., 2012). Many researchers have begun using participatory and collaborative design methods with children on the autism spectrum.. Porayska-Omosta et al. (2012) attribute the increased interest in the CBR approach in this area to several reasons, including the children's interest in technology and a dramatic increase in awareness around the prevalence of autism.

Given the technological background and interest that is growing in Kenya and other East African countries in the region (described in the beginning of this section), there are many possibilities for research and development in the area of digital assistive technologies in the region.

Present: A Focused Look at Special Education Schools in the Kisumu region

In the summer of 2014, our collaborative team engaged in two forms of consultations: school visits and a broad stakeholder meeting. We visited three special education institutions: Joyland Special School for the Physically Handicapped (Kisumu); Lutheran Church Special School for the Mentally Challenged (Kisumu); and St. Francis Oriang Community School (Kombewa). These activities were not formal research but rather a set of community consultations to explore the interest in and need for future research. The goal of this step was to gain preliminary first-hand knowledge about the current use of digital assistive technology in a variety of schools.

- **Joyland Special School for the Physically Handicapped:** Joyland Special School for the Physically Handicapped is a government-run school in Kisumu that provides primary and secondary school education to 169 students (56% boys). Students mainly have motor disabilities but a few students have low-vision and speech difficulties. There are several students with epilepsy and several who are affected by AIDs and chronic ailments. Some students are refugees from Sudan and Somalia. There were 15 teachers and several teacher trainees. The school receives partial funding for essential staff, AT (non-digital) and learning material. The school specifically caters to children with physical disabilities and uses a regular curriculum.

- **St. Francis Oriang Community School:** St. Francis Oriang Community School is a community-based school located in Seme Sub-County (about an hour from Kisumu City). At the time of our visit, this school was privately funded; it has since become a government-run school. This school employs an inclusive education model where children (mainly with physical disabilities) attend the same classes and program as local children without disabilities. Out of 275 students, 56 have special needs. Out of 12 teachers at the school, three have additional training in special education needs and there are two computer instructors with specialized education in the government's computer learning program.
- **Lutheran Church Special School for the Mentally Challenged:** The Lutheran Church Special School for Mentally Challenged provides educational services to children with various cognitive disabilities including autism and various degrees of developmental and learning disabilities. There are 114 students (37% boys) with most boarding at the school (Cohen, 2012). There are 22 teachers for the students, of whom eight are trained in computer instruction. Students are provided with academic and vocational skills, with more emphasis on vocational skills that could possibly make them more independent in the community.

In each site, we visited all relevant school facilities, including computer rooms, classrooms and, where they existed, dormitories. We met with school principals, teachers of computer science, and in two of the schools (Joyland and St. Francis Oriang), had a brief opportunity to observe and interact with students in their daily routines although not in the computer classrooms

The range of schools visited, including urban and rural and state sponsored and privately funded, provided an overview of different approaches to special education is practiced in the region. While the approach of each school was different, all the schools were equipped with desktop computers in their computer classrooms. We identified several elements that present potentials and challenges that can be explored in future interventions.

Aggregated Observations

- **Availability of Technology:** All schools have a computer lab setup and there are staff and teachers who are familiar with computers. Figure 1 shows the computer labs in these schools. The skills of the instructors range from ability to do basic computer maintenance and setup to advanced computer programming. Most computers are Windows-based and are older donated models but most are usable and in working condition. While it is possible to connect to the Internet using standard USB dongles, the schools have identified alternative off-line multimedia resources (e.g., MS-Encarta) that lower connection delays and cost. One school (St. Francis Oriang Community School) had used excess computers to create an Internet Café on the premises and provided services such as Internet access and printing to the broader community. This school is located in a rural area and thus can provide a valuable resource for the community outside the school operation hours. While there is potential for a conflict of interest between the needs of school children and community members who would pay to use the computers, this setup also provides an opportunity for children-run technology centers that might serve as a vocational training resource as well.

  At the Joyland Special School, the computer curriculum follows that specified by the education board. It covers the basic use of computers and applications (including a lot of conceptual and hands-off information) and continues to introduce basics of programming in later years (using VB). Most classes would have students working in teams of 4 per computer. At the other schools, we did not discuss the computer instruction curriculum.
- **Inclusive Education:** Inclusive Education (IE), as defined as a human right by the United Nations, is the ambition and goal that all children regardless of ability or disability should be educated in mainstream classes in their district schools, receiving education and support as required by their needs (UN, 2006). With the exception of the Lutheran Church Special School for Mentally Challenged, the schools employed an inclusive education model where students with disabilities attend the same classes as students without disabilities. This is to prepare both populations for future social integration. In the case of students with mental health problems or cognitive disabilities, while there was emphasis on helping them become more independent in their private lives (e.g., through hygiene and basic communication training), expectations are low as to their potential inclusion in society. The underlying reasons for these expectations, including the effect of accepting children late into programs, are discussed in a previous report (Cohen, 2012).

- **Physical Disabilities, Cognitive Disabilities and Mental Health:** Throughout our stay we observed a fundamental difference in approaches to physical disabilities and cognitive disabilities. While physical disability is accepted as a condition that can be alleviated using education, technology and community support, cognitive disability still poses many challenges, not only in terms of the tools or processes needed but also in terms of public perception. There is a great lack of resources for children and adults with cognitive disabilities. There are specific programs to help these children with basic self-care and some vocational training (sewing, textile making) but, as noted above, expectations are low (Cohen, 2012). However, there was also evidence that having hands-on activities with a clear outcome can be beneficial for the children. Figure 2 shows a textile-making workshop at the Lutheran Church Special School.
- **Lack of Digital Assistive Technology:** None of the schools had any specific assistive software or hardware available. While the computers are there, it would be hard to imagine how children with motor, cognitive or communication disabilities can use them effectively without specific and appropriate support. However, perhaps more important than the existence of computers and technology in the schools, was the prevalence of a positive attitude towards technology and a recognition of its potential to help children with disability (and others). This atmosphere of acceptance could be an important resource in deploying projects in this region in the future.

Present: Stakeholder Meeting

Participants in the meeting included government representatives from the Department of Social Services, Department of Children, County Office, NEMA (National Environment Management Authority) and the Ministry of Agriculture; NGO representatives from Family AIDS Care and Education Services (FACES), Association for the Physically Disabled of Kenya (APDK), Support for Tropical Initiatives in Poverty Alleviation (STIPA), the National Council for People Living with Disabilities (NCPWD),and Partners in Transformation (PCT).

Figure 3 shows a network map created during the meeting by the stakeholders to illustrate existing, potential and desired relationships between the different stakeholders. As is evident from the map, the Ministries of Education, Labor, and Agriculture, the NCPWD, and APDK are all strongly connected to one another and act as hubs for a web of connections to many other organizations, while other organizations have only a few connections. This map was useful for identifying not only where connections could be strengthened but also where relationships could be leveraged to address gaps in accessibility.

Figure 1. Computer facilities in Kisumu special education schools: (Clockwise starting from top-left) Lutheran Church Special School for Mentally Challenged, St. Francis Oriang Community School and Joyland Special School for the Physically Handicapped

Figure 2. Vocational training workshop at Lutheran Church Special School for Mentally Challenged

Figure 3. Network map created at the stakeholders meeting

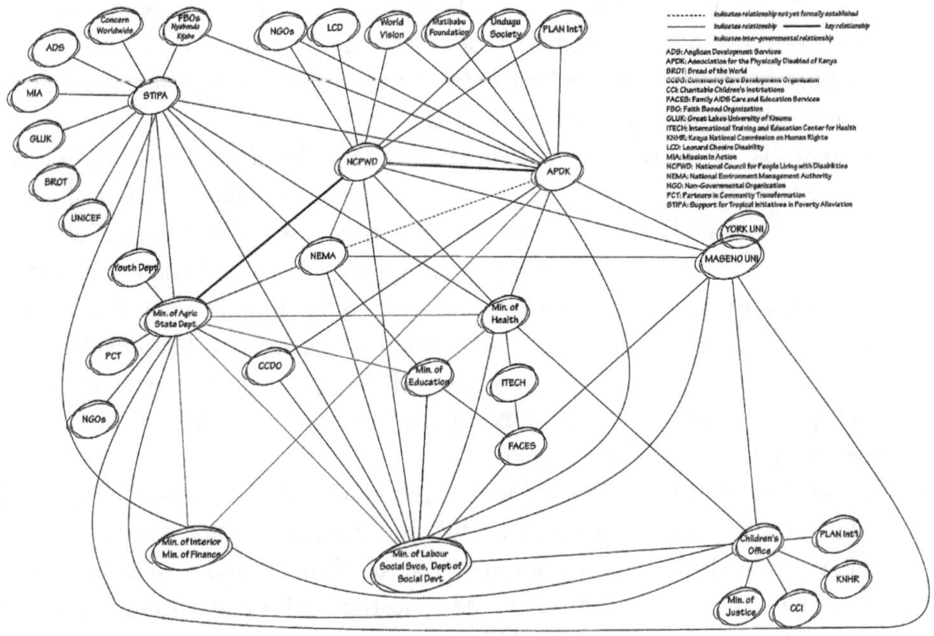

Several potential directions for development were identified. These included: public-private partnerships, inclusive education, routes of e-citizenship for people with disabilities, creation of employment opportunities for people with disabilities, and improved resources for people with mental health and cognitive disabilities. Dominant themes are described in more detail below.

Dominant Themes of Stakeholders Meeting

- **Interest in Government-Industry Collaboration:** As mentioned in the beginning of this section, in the past, successful public-private partnerships have been formed to provide marginalized populations with technology, including children with visual impairments and children living in slums. Many stakeholders identified the success of ICT business incubation centers such the iHub and m:Lab as places to seek industry partners and expertise.
- **Potentials of E-Citizenship:** Kenya, through the homegrown development and use of crowd-sourcing e-citizenship tools such as Ushahidi, is also a leader in e-citizenship initiatives. There are possibilities to customize and use Ushahidi and similar platforms, for example the South African, open-source SMS platform FrontlineSMS (http://www.frontlinesms.com), for use by people with disabilities.

- **Inclusive Education:** Away from the traditional approaches of establishing special schools for children with disabilities, the stakeholders emphasized the need to strengthen integration in learning institutions, a move that would hasten rehabilitation of children with special needs, bring about more community involvement and reduce stigma.
- **Potentials of E-Work for People with Disabilities:** The stakeholders suggested exploring possibilities of employment for people with disabilities as an important direction. A recognized barrier was the high cost of existing and imported AT (e.g., the JAWS solution for people with blindness and low-vision). Finding solutions that are relevant to the Kenyan context and with low acquisition barriers (i.e., cost) would be necessary. The development of homegrown technology such as m-Pesa for making payments using cell phones and microwork opportunities using cell phone and distributed technology have inspired these ideas.
- **Potentials of AT for People with Mental Health Issues and Cognitive Disabilities:** The stakeholders at this meeting also acknowledged the lack of resources for children with cognitive disabilities and stated that while there have been improvements to the education and social systems' response to children with physical disabilities, the case for children with cognitive disabilities is different. Exploring possible tools or technologies to help this population have better access to independent living, education, and perhaps even employment was identified as important.

Future: Potential Direction for Exploration

Given the fertile technological and academic setting in Kenya described at the beginning of this section, and observations from fieldwork at the schools, we have outlined several aspects of an envisioned direction for developing a program to sustainably design and deploy digital AT that would help the move towards the goal of providing inclusive special education to Kenyan children.

We believe a promising direction is to establish a community-university engagement initiative strengthened through collaboration with an international university with experience and expertise in development work and AT development. Similar to CMU's initiative in Rwanda (www.cmu.edu/rwanda/) but with a focus on AT, a joint interdisciplinary graduate program in digital technology and design can be conducted where graduate and advanced undergraduate Computer Science, Social Work and Special Needs Education students design, implement and deploy digital AT. Situating the course in the university setting allows for the undertaking of relevant research and development methods that provide an evidence-based approach to technology development and also provide opportunities to share and disseminate knowledge effectively. Being part of an ongoing university program would also ensure sustainability of activities and continued interest in the initiative through ongoing technology development and the training of local students both for practical placements in the community and as future service providers. Participatory design approaches would also provide avenues for community–based social interventions that promote social inclusion for children with disability. Where these approaches have been established like iHub (http://ihub.co.ke/) in Nairobi, they have succeeded in creating spaces for innovative and personalized digital assistive technologies by young developers and people around them thus addressing the multiple social and individual aspects of stigma that result in social exclusion.

As described previously, some promising implementation directions in the past have included the use of free open source software and the inclusion of public private partnerships to deploy the technology to a larger user group. The program would employ these elements and explore ways that FOSS can be used to implement new systems. It could also collaborate with industry partners to possibly commercialize some of the research outcomes. In recent years, the field of HCI has taken a more active role in bringing about societal benefits through adopting democratic and collaborative design and research approaches, such as Action Research and Community Based Research (Hayes, 2011). Participatory Action Research methodologies that involve design "in the wild" (i.e., in situ and in the actual context in which they are to be used) and co-design with close collaboration with NGOs and schools seem most promising (Rogers & Marsden, 2013; Gitau & Marsden, 2009; Marsden, 2008).

We believe that such a program has the potential to contribute to the growth of the next generation of AT specialists, not only in Kenya but also in East Africa. Conducting the course locally will ensure the program's sustainability and relevance, personnel and institutional development as well as knowledge transfer, skill and development. Additionally, using-state-of-the-art technologies and dissemination methods would support the creation of new links between institutions and organizations in Kenya and the international community.

CONCLUSION

In this chapter, we have reviewed the relevant research literature on ICT4D, PD of AT and Critical Disability Studies and have grounded the existing theory with preliminary observations from a field trip to a range of special education schools in the Kisumu region of Kenya. Additionally, we have shared ideas from a stakeholder forum discussing limitations and future direction for AT in the region. The landscape of AT development and deployment is full of possibilities and challenges in Kenya. On the one hand, Kenya's new wave of tech-savvy and creative individuals and organizations can bring about positive change through innovative ideas sensitive to the Kenyan cultural and societal context with translation possibilities for other contexts beyond Kenya or East Africa. On the other hand, difficulties arising from the current lack of existing ATs, of access to tools and expertise in the design and deployment of new solutions and challenges in regional and national coordination present barriers to breakthroughs in inclusive education. These barriers can be removed through innovative programs that engage NGOs, universities, ICT businesses and most importantly, people with disabilities and their caregivers, teachers and families in participatory and inclusive design processes. Our goal is that this chapter provides background and motivation for such projects in the future.

REFERENCES

Anthony, L., Prasad, S., Hurst, A., & Kuber, R. (2012).A participatory design workshop on accessible apps and games with students with learning differences. *Proceedings of the 14th International ACM SIGACCESS Conference on Computers and Accessibility*, 253-254. doi:10.1145/2384916.2384979

Benton, L., Johnson, H., Brosnan, M., Ashwin, E., & Grawemeyer, B. (2011). IDEAS: an interface design experience for the autistic spectrum.*Proceedings of the 2011 annual conference extended abstracts on Human factors in computing systems*, 1759-1764. doi:10.1145/1979742.1979841

Benton, L., Vasalou, A., Khaled, R., Johnson, H., & Gooch, D. (2014). Diversity for design: a framework for involving neurodiverse children in the technology design process.*Proceedings of the 2014 annual ACM conference on Human factors in computing systems*, 3747-3756. doi:10.1145/2556288.2557244

Bernardini, S., Porayska-Pomsta, K., & Smith, T. J. (2014). ECHOES: An intelligent serious game for fostering social communication in children with autism. *Information Science*, *264*, 41–60. doi:10.1016/j.ins.2013.10.027

Bezy, M. (2011). Carnegie Mellon University, First US Research Institution to Open a Campus In Africa. *Africa Oye*. Retrieved from http://brel54.blogspot.com.es/2011/09/carnegie-mellon-university-first-us.html

Bezy, M. (2013). Striving to Become Africa's First Silicon Valley: Comparing Kenya and Rwanda. *Africa Oye*. Retrieved from http://brel54.blogspot.com.es/2013/03/striving-to-become-africas-first.html

Botelho, F. (2009). Public-Private Partnerships for Low and No-Cost Assistive Technologies. *ITU Workshop on Global Internet Access for Persons with Disabilities*. Retrieved from https://www.itu.int/dms_pub/itu-t/oth/06/27/T06270000060042PDFE.pdf

Boyer, E. L. (1990). *Scholarship Reconsidered: Priorities of the Professoriate*. Carnegie Foundation for the Advancement of Teaching.

Brederode, B., Markopoulos, P., Gielen, M., Vermeeren, A., & de Ridder, H. (2005). pOwerball: the design of a novel mixed-reality game for children with mixed abilities. *Proceedings of the 2005 conference on Interaction design and children(IDC'05)*, 32–39. doi:10.1145/1109540.1109545

Chepken, C., Mugwanya, R., Blake, E., & Marsden, G. (2012). ICTD interventions: Trends over the last decade.*Proceedings of the Fifth International Conference on Information and Communication Technologies and Development*, 241-248. doi:10.1145/2160673.2160704

Chernikova, E. (2011). *Shoulder to shoulder or face to face? Canada's university-civil society collaborations on research and knowledge for international development*. Ottawa: International Development Research Centre.

Cohen, T. (2012). *The Invisible Disability: Perceptions and Potential of Children with Autism in Kisumu, Kenya*. Independent Study Project (ISP) Collection. Paper 1245.

Corker, M., & Shakespeare, T. (Eds.). (2002). *Disability/postmodernity: Embodying disability theory*. Bloomsbury Publishing.

Cullen, D. (2006). JAWS and Dolphins for Nairobi's blind students. *The Register.* Retrieved from http://www.theregister.co.uk/2006/03/29/computer_aid_kilimani_primary_school/

Davidson, J. L., & Jensen, C. (2013). What health topics older adults want to track: a participatory design study.*Proceedings of the 15th International ACM SIGACCESS Conference on Computers and Accessibility*, 1-8. doi:10.1145/2513383.2513451

Dawe, M. (2007). Let me show you what I want: engaging individuals with cognitive disabilities and their families in design. *Proceedings of the annual ACM conference on Human factors in computing systems (CHI'07)*, 2177-2182. doi:10.1145/1240866.1240976

Dodson, L. L., Revi Sterling, S., & Bennett, J. K. (2013). Minding the gaps: cultural, technical and gender-based barriers to mobile use in oral-language Berber communities in Morocco.*Proceedings of the 6th International Conference on Information and Communication Technologies and Development*, 79-88. doi:10.1145/2516604.2516626

Farell, G. (2007). Survey of ICT and Education in Africa: Rwanda Country Report. Washington, DC: World Bank. Available from https://openknowledge.worldbank.org/handle/10986/9415

Frauenberger, C., Good, J., & Alcorn, A. (2012). Challenges, opportunities and future perspectives in including children with disabilities in the design of interactive technology. *Proceedings of the 11th International Conference on Interaction Design and Children*, 367-370. doi:10.1145/2307096.2307171

Frauenberger, C., Good, J., & Keay-Bright, W. (2010).Phenomenology, a framework for participatory design. *Proceedings of the 11th Biennial Participatory Design Conference*, 187-190. doi:10.1145/1900441.1900474

Frauenberger, C., Good, J., & Keay-Bright, W. (2011). Designing technology for children with special needs: Bridging perspectives through participatory design. *CoDesign*, *7*(1), 1–28. doi:10.1080/15710882.2011.587013

Galis, V. (2011). Enacting disability: How can science and technology studies inform disability studies? *Disability & Society*, *26*(7), 825–838. doi:10.1080/09687599.2011.618737

Galliers, J., Wilson, S., Roper, A., Cocks, N., Marshall, J., Muscroft, S., & Pring, T. (2012). Words are not enough: empowering people with aphasia in the design process.*Proceedings of the 12th Participatory Design Conference*, 51-60. doi:10.1145/2347635.2347643

Gitau, S., &Marsden, G. (2009).Fair Partnerships–Working With NGOs. *Human-Computer Interaction–INTERACT*. Springer Berlin Heidelberg.

Goldenberg, E. P. (1979). *Special Technology for Special Children*. Baltimore, MD: University Park Press.

Gowans, G., Campbell, J., Alm, N., Dye, R., Astell, A., & Ellis, M. (2004). Designing a multimedia conversation aid for reminiscence therapy in dementia care environments. *Proceedings of the 2004 annual conference extended abstracts on Human factors in computing systems*, 825-836. doi:10.1145/985921.985943

Greenbaum, J., & Kyng, M. (Eds.). (1991). *Design at Work: Cooperative Design of Computer Systems*. Lawrence Erlbaum Associates.

Grönlund, Å., Lim, N., & Larsson, H. (2010). Effective use of assistive technologies for inclusive education in developing countries: Issues and challenges from two case studies. *International Journal of Education and Development Using ICT*, *6*(4), 5-26.

Harrison, S., Tatar, D., & Sengers, P. (2007). The three paradigms of HCI. *Alt. Chi. Session at the SIGCHI Conference on Human Factors in Computing Systems*, 1-18.

Hayes, G. R. (2011). The relationship of action research to human-computer interaction. *ACM Transactions on Computer-Human Interaction*, *18*(3), 15. doi:10.1145/1993060.1993065

Horowitz, C. R., Robinson, M., & Seifer, S. (2009). Community-based participatory research from the margin to the mainstream: Are researchers prepared? *Circulation*, *119*(19), 2633–2642. doi:10.1161/CIRCULATIONAHA.107.729863 PMID:19451365

Hourcade, J. P., Garzotto, F., Rozga, A., Tentori, M. E., Markopoulos, P., Pares, N., Good, J., Pain, H., &Alper, M. (2014). Supporting children with complex communication needs. *CHI '14 Extended Abstracts on Human Factors in Computing Systems*, 119-122.

Hurst, A., & Kane, S. (2013). Making making accessible.*Proceedings of the 12th International Conference on Interaction Design and Children*, 635-638. doi:10.1145/2485760.2485883

Israel, B. A., Schulz, A. J., Parker, E. A., & Becker, A. B. (1998). Review of community-based research: Assessing partnership approaches to improve public health. *Annual Review of Public Health*, *19*(1), 173–202. doi:10.1146/annurev.publhealth.19.1.173 PMID:9611617

IST-Africa. (2016). *Report on ICT Initiatives and Research Capacity in IST-Africa Partner Countries*. Available from: http://www.ist-africa.org/home/files/IST-Africa_ICTInitiatives_ResearchCapacity_v2_310116.pdf

Jacobs, S. (1999). *Section 255 of the Telecommunications Act of 1996: Fueling the Creation of New Electronic Curbcuts*. Retrieved from http://www.accessiblesociety.org/topics/technology/eleccurbcut.htm

Kassam, K. A., & Tettey, W. J. (2003). Academics as citizens: Collaborative applied interdisciplinary research in the service of communities. *Canadian Journal of Development Studies*, *24*(1), 155–174. doi:10.1080/02255189.2003.9668902

Keay-Bright, W. E. (2007). The Reactive Colours Project: Demonstrating Participatory and Collaborative Design Methods for the Creation of Software for Autistic Children. *Design Principles and Practices: An International Journal*, *1*(2), 7–16. doi:10.18848/1833-1874/CGP/v01i02/37623

Kensing, F., &Blomberg, J. (1998). Participatory design: Issues and concerns. *Proceedings of Computer Supported Cooperative Work*, 167-185.

Kientz, J. A., Hayes, G. R., Westeyn, T. L., Starner, T., & Abowd, G. D. (2007). Pervasive Computing and Autism: Assisting Caregivers of Children with Special Needs. *IEEE Pervasive Computing Magazine*, *6*(1), 28–35. doi:10.1109/MPRV.2007.18

Mankoff, J., Hayes, G. R., & Kasnitz, D. (2010). Disability studies as a source of critical inquiry for the field of assistive technology.*Proceedings of the 12th international ACM SIGACCESS conference on Computers and accessibility*3-10. doi:10.1145/1878803.1878807

Marsden, G. (2008). Toward Empowered Design. *IEEE Computer*, *41*(6), 42–46. doi:10.1109/MC.2008.207

Marsden, G., Maunder, A., &Parker, M. (2008). People are people, but technology is not technology. *Philosophical Transactions of the Royal Society A: Mathematical, Physical and Engineering Sciences*, *366*(1881), 3795-3804.

Matinde, V. (2015). *Re-looking at Kenya's digital learning school project*. IDG Connect. Available online: http://www.idgconnect.com/blog-abstract/9273/re-looking-kenya-digital-learning-school-project

Mbuvi, D. (2012). Kenya ICT Board and CMU launch international software standard. *PC Advisor*. Retrieved from http://www.pcadvisor.co.uk/news/software/3336503/kenya-ict-board-cmu-launch-international-software-standard/

Millen, L., Cobb, S., & Patel, H. (2011). A method for involving children with autism in design. *Proceedings of the 10th International Conference on Interaction De sign and Children(IDC '11)*,185-188. doi:10.1145/1999030.1999057

Minkler, M. (2005). Community-based research partnerships: Challenges and opportunities. *Journal of Urban Health*, *82*(2), ii3–ii12. PMID:15888635

Moffatt, K., McGrenere, J., Purves, B., & Klawe, M. (2004). The participatory design of a sound and image enhanced daily planner for people with aphasia.*Proceedings of the 2004 annual ACM conference on Human factors in computing systems(CHI'04)*, 407-414. doi:10.1145/985692.985744

Mulupi, D. (2013). *Kenya's tech industry: over-hyped or just learning to walk? How we made it in Africa.* Retrieved from http://www.howwemadeitinafrica.com/kenyas-tech-industry-over-hyped-or-just-learning-to-walk/

National Coordinating Agency for Population and Development (NCPD). (2008). *Kenya National Survey for Persons with Disabilities – Preliminary Report (KNSPWD).* National Coordinating Agency for Population and Development.

Oduor, E., Neustaedter, C., Judge, T., Hennessy, K., Pang, C., & Hillman, S. (2014). How technology supports family communication in rural, suburban, and urban Kenya. *Proceedings of the 32nd annual ACM conference on Human factors in computing systems (CHI '14)*, 2705-2714. doi:10.1145/2556288.2557277

Oliver, M. (1998). Theories of disability in health practice and research. *British Medical Journal, 317*(7170), 1446–1449. doi:10.1136/bmj.317.7170.1446 PMID:9822407

Piper, A. M., O'Brien, E., Morris, M. R., & Winograd, T. (2006). SIDES: a cooperative tabletop computer game for social skills development.*Proceedings of the 2006 20th anniversary conference on Computer supported cooperative work*, 1-10. doi:10.1145/1180875.1180877

Porayska-Pomsta, K., Frauenberger, C., Pain, H., Rajendran, G., Smith, T., Menzies, R., & Lemon, O. et al. (2012). Developing technology for autism: An interdisciplinary approach. *Personal and Ubiquitous Computing, 16*(2), 117–127. doi:10.1007/s00779-011-0384-2

Preiser, W. F. (2001). *Universal design handbook*. New York: McGraw-Hill.

Reed, P., & Bowser, G. (2005). Assistive technologies and the IEP. In D. Edyburn, K. Higgins, & R. Boone (Eds.), Handbook of Special Education Technology Research and Practice. Academic Press.

Rogers, Y. (2011). Interaction design gone wild: Striving for wild theory. *Interaction, 18*(4), 58–62. doi:10.1145/1978822.1978834

Rogers, Y., & Marsden, G. (2013). Does he take sugar? Moving beyond the rhetoric of compassion. *Interaction, 20*(4), 48–57. doi:10.1145/2486227.2486238

Rosenbaum, P., & Stewart, D. (2004). The World Health Organization international classification of functioning, disability and health: A model to guide clinical thinking, practice, and research in the field of cerebral palsy. *Seminars in Pediatric Neurology, 11*(1), 5–10. doi:10.1016/j.spen.2004.01.002 PMID:15132248

Sanders, E. B. N., & Stappers, P. J. (2008). Co-creation and the new landscapes of design. *CoDesign*, *4*(1), 5–18. doi:10.1080/15710880701875068

Schuler, D., & Namioka, A. (Eds.). (1993). *Participatory Design: Principles and Practices*. Hillsdale, NJ: Lawrence Erlbaum Associates.

Sika, V., Sambuli, N., Orwa, A., & Salim, A. (2014). ICT and Governance in East Africa: A Landscape Analysis in Kenya, Uganda and Tanzania. iHub Research.

Simmons, D. (2014). Keepod: Can a $7 stick provide billions computer access? *BBC Click*. Retrieved from http://www.bbc.com/news/technology-27346567

Smith, D. (2012). Kenya building a digital future in Africa's silicon savannah. *The Guardian*. Retrieved from http://www.theguardian.com/world/2012/oct/30/kenya-silicon-savannah-digital-technology

Suarez-Balcazar, Y., Harper, G. W., & Lewis, R. (2005). An interactive and contextual model of community-university collaborations for research and action. *Health Education & Behavior*, *32*(1), 84–101. doi:10.1177/1090198104269512 PMID:15642756

Treviranus, J. (2014). Leveraging the Web as a Platform for Economic Inclusion. *Behavioral Sciences & the Law*, *32*(1), 94–103. doi:10.1002/bsl.2105 PMID:24549724

UNESCO. (2007). *The UNESCO ICT in Education Programme, Bangkok*. Available from: http://unesdoc.unesco.org/images/0015/001567/156769e.pdf

United Nations (UN). (2006). *Convention on the Rights of Persons with Disabilities*. United Nations. Retrieved from http://www.un.org/disabilities/convention/conventionfull.shtml

Unwin, P. T. H. (Ed.). (2009). *ICT4D: Information and Communication Technology for Development*. Cambridge University Press.

Visser, F. S., Stappers, P. J., Van der Lugt, R., & Sanders, E. B. (2005). Contextmapping: Experiences from practice. *CoDesign*, *1*(2), 119–149. doi:10.1080/15710880500135987

Wendell, S., & Linton, S. (2000). The rejected body: Feminist philosophical reflections on disability. *Signs (Chicago, Ill.)*, *25*, 587–589.

Wilson, S., Roper, A., Marshall, J., Galliers, J., Devane, N., Booth, T., & Woolf, C. (2015). Codesign for people with aphasia through tangible design languages. *CoDesign*, *11*(1), 21–34. doi:10.1080/15710882.2014.997744

World Health Organization (WHO). (2001). *International Classification of Functioning, Disability and Health (ICF)*. World Health Organization.

World Health Organization (WHO). (2009). *Assistive devices/technologies*. Retrieved from http://www.who.int/disabilities/technology/en/

Wu, M., Richards, B., & Baecker, R. (2004). Participatory design with individuals who have amnesia. *Proceedings of the 8th Participatory Design Conference*, 214-223. doi:10.1145/1011870.1011895

Chapter 7
Integrating Digital Literacies Into an Undergraduate Course:
Inclusiveness Through Use of ICTs

Meke I. Kapepo
University of Cape Town, South Africa

Tabisa Mayisela
University of Cape Town, South Africa

ABSTRACT

The purpose of this chapter is to share a successful case of how ICT was used for educational inclusiveness by students from previously disadvantaged backgrounds. This chapter further illuminates how students could be supported in acquiring digital and academic literacy skills, within a discipline context. Lecturers often find it difficult to define digital literacy beyond the use of ICT in a classroom, or understand how students' acquisition of digital literacy skills could be supported or further expanded into students' academic literacy skills. This piece of work therefore, provides insight to academics, researchers and students, on how: (1) ICT could be integrated into the curriculum; (2) digital literacy could be embedded into the curriculum; and (3) students could be supported in acquiring graduate attributes and skills that make them fit for the workplace.

DOI: 10.4018/978-1-5225-2565-3.ch007

INTRODUCTION

The ubiquitous use of Information and Communication Technology (ICT) in the knowledge-driven business world, poses a dire need of graduates with the necessary literacies that deem them fit for such workplaces. It is therefore important to investigate how higher education could possibly equip students with these literacies, particularly, on how the curriculum could be designed to potentially foster these literacies. ICT integration into the curriculum is an overarching phenomenon in higher education, and literature on this demonstrates the shifting in meaning of integration, ranging from the use of computers and networks as tools for teaching and learning, to the appropriation of ICT to enhance different pedagogical approaches and student learning experiences (Venkatesh, Croteau, & Rabah, 2014). Recent studies further demonstrate how emerging technologies are being used in South Africa to enhance student learning in higher education (Bharuthram & Kies, 2013; Bozalek et al., 2013; Cloete, 2014; Czerniewicz & Brown, 2013). Irrespective of ICT being integrated into the curriculum, there is limited literature on how higher education fosters the development of students' digital literacies that are required for learning and subsequently, the workplace. Bozalek et al (2014) confirm that there is a disjuncture between curriculum design and what is required of working professionals, hence a recommendation to incorporate authentic tasks to promote contextually relevant learning. It is in this context that educational researchers have increasingly motivated for the integration of these digital literacies into the curriculum (Kirkwood, 2006; Beetham et al. 2009). Brown and Mayisela (2015) suggest that one of the possible means of addressing this, is by having Higher Education Institutions (HEIs) broadening their conceptualization of digital literacy beyond computer or technology literacy because the concept of digital literacy "encompasses a range of practices, including computer literacy, information literacy, media literacy, communication literacy, visual literacy and technology literacy" (p.16).

According to the Joint Information Systems Committee (JISC, 2014), digital literacies are "those capabilities, which fit an individual for living, learning and working in a digital society" (JISC, 2014). The Open University, UK, describes digital literacy to include the "ability to find and use information (otherwise known as information literacy) but goes beyond this to encompass communication, collaboration and teamwork, social awareness in the digital environment, understanding of e-safety and creation of new information" (Open University, 2012). Integrating digital literacies into the curriculum therefore, implies that the curriculum is more likely to foster authentic and multimodal learning experiences (Chase & Laufenberg, 2011). The literature suggests that students have a greater opportunity to take control of their own learning, and are better able to acquire academic and digital literacies iteratively, when these are embedded in practical tasks made available through authentic projects. For instance, related literature (such as Kirkwood, 2006) on developing students' digital information literacy skills suggest that these skills be developed in the context of the course or discipline because that allows students an opportunity to learn how to "frame pertinent questions with which to evaluate and select appropriate sources" for the course. Furthermore, research conducted by Ng (2012) on teaching digital literacy to pre-service teachers at a university in Australia revealed that students acquired technology and cognitive skills, and also created useful artefacts that they could use for their future teaching when placed in schools. As Brown and Mayisela (2015) noted; there are few emerging pockets of work around digital literacy in some South African HEIs. Hence, there may be limited research on the integration of digital literacies into the curriculum in South African HEIs. The purpose of this chapter is to make a contribution to this effect.

An approach of integrating digital literacies into an undergraduate curriculum to develop and support students' academic and digital literacy skills is presented. It does so by discussing how ICT can be used to enhance educational inclusiveness of first year higher education students who hail from predominantly less privileged socio-economic backgrounds. A project based authentic task was developed to teach e-commerce concepts to a group of undergraduate students in the faculty of Commerce. The course provides an introduction to information systems where information systems are described from a business perspective. Teaching electronic commerce concepts to students with limited knowledge and almost no prior exposure to ICTs is challenging because at times students cannot relate to theory concepts explained in class. These challenges stem from students' backgrounds and their education (with less-resourced schools) from the marginalized sections of society. This curriculum intervention demonstrates how the socio-economic barriers that lead to educational exclusion could be broken down, by integrating ICT into the curriculum and further supporting students as they learn in these digital contexts.

The structure of this chapter is as follows - Section one provides a background on ICT use in higher education and views of how on digital literacies can be integrated into the curriculum. Student digital inequalities are discussed to highlight the gap in literature and point out challenges in developing contextualized and inclusive curriculum in higher education. The next section discusses the authentic task by reporting on how the authentic learning pedagogical approach was used to develop a project-based task to integrate digital literacies into the curriculum. This is discussed by demonstrating how the Digital Information Literacy (DIL) framework was used to design the project and how we applied the compete areas to evaluate the skills students acquired and being able to link them to their career goals. Finally, the chapter concludes by pointing out future directions for the potential use of ICTs to enhance authentic learning for inclusive education. escribe the general perspective of the chapter.

BACKGROUND

According to (Bélisle, 2006), "the global socio-economic transformation we are experiencing today has brought information, knowledge and innovation as the main source of economic growth and employment opportunities" (2006:51). These three elements are crucial for the development of new products and solving socio-economic problems in any country (Bélisle, 2006). Businesses, as presented in this case, are no exception; they need these elements for their business processes. Business processes include, accessing, processing, and producing information and products, and other business transactions. However, worldwide, graduate employability has become a concern (Haywood, Haywood & Macleod, 2006; UNESCO, 2014), more particularly in the developing countries (Lau, 2006). This is due to the graduates' lack of the essential literacies required in the workplace. UNESCO (2014) also reports that graduates from developing countries lack digital and information literacy skills that are required for participation in the 21st century knowledge-based economies. This challenge therefore, calls to equip graduates with essential digital literacies necessary to carry out workplace functions, including business processes. Supporting students to develop 'learning literacies for the digital age', (Beetham, McGill and Littlejohn, 2009:8), presupposes that students may be better prepared for working in the knowledge-based economy where ICT and digital forms of information communication and management prevail.

In the context of this case study, an authentic task is used to help students acquire learning literacies as well as digital literacies for the e-business environment. An authentic task is a project based activity that emphasizes understanding and application of the task at hand. The authentic task served as a means to introduce students to the socio-cultural environment of e-business processes, by providing students with a shopping experience on an e-commerce platform.

This study was conducted with a group of first year students studying information systems in the faculty of commerce. Information systems is a service course offered to bachelor of commerce (B.Com) or Business Science degree (BusSci) students specializing in Economics, Accounting and Actuarial science. Given the nature of the service course, about 70% of the students in this group come without prior knowledge and access to digital technologies. Student digital inequalities are discussed in the next section.

MAIN FOCUS OF THE CHAPTER

Student Digital Inequalities

Due to the ubiquitous use of information and communication technology (ICT) in the digital, knowledge-driven business world, it becomes very important that higher education prepares and equips students with the necessary skills for such workplaces. However, literature illustrates that students who enter HEIs have diverse educational, cultural, economic and more particularly, digital backgrounds. On one hand, are students who have had access to, and extensive experience in using digital media, since the early ages of their lives. Prensky (2001), refers to these as "digital natives". On the other hand, are those who lack ICT experience or have barely used a computer prior university. Czerniewicz and Brown (2013) refer to these students as "digital strangers".

A number of studies confirm the digital inequalities of students who enroll in South African HEIs (Czerniewicz & Brown, 2013; Fisher, 2011). For instance, Oyedemi (2012) confirms these inequalities in his study where he surveyed students from ten South African universities. Findings from his study revealed a pattern of digital inequalities that further reproduce existing structures of social inequalities in South Africa (Oyedemi, 2012). Another study conducted at a South African university, to elicit first year students' in-school and out-of-school digital literacy practices, revealed that out from a total of twenty participants (16 from various regions of South Africa, two from Southern African Development Community countries, one each from China and Iraq), fifteen of the S.A and one of the SADC students had no access to digital technology at school (Kajee & Balfour, 2011), one of them had a computer at home and one accessed the Internet from an Internet Cafés closer to his home (Kajee & Balfour, 2011).

Students enrolled for this study come from educational backgrounds that have not adequately prepared them for university. The digital inequalities prevalent in this group of students were lack of access to computers, internet and skills to utilize them. To further explore these digital inequalities, makes capacitating them with technology skills, at the least, a daunting task. Teaching and learning with ICTs in such a diverse class may further exacerbate the digital divide.

Given the inequalities in digital access, skills and literacy, this calls for the higher education sector to investigate how they could use ICT to bridge this divide, particularly ways of ensuring inclusivity of the previously disadvantaged students. As a means of response, this study examined how ICT can be integrated into the curriculum to develop students' digital literacy skills and other necessary skills required for the workplace. Educational researchers have increasingly motivated for the integration of these literacies into the curriculum (Beetham et al. 2009; Brown

2012). In this instance, the nature of teaching and learning tends to change as the digitally mediated environment enables learning practices that are multimodal and authentic (Chase & Laufenberg 2011). The literature suggests that students have a greater opportunity to take control of their own learning, and are better able to acquire academic and digital literacies iteratively, when these are embedded in practical tasks made available through authentic projects. In an increasingly digitally mediated world, students require digital literacy skills to take advantage of new work and learning environments and to become independent and collaborative graduates. It is therefore important to integrate these literacies in the curriculum to ensure students attain these skills.

The Task

A class of 48 students was divided into groups of 8-10 students. The purpose of grouping the students was to encourage them to collaboratively discuss and learn the online ordering process. The lecturer identified an e-commerce business and bought vouchers that were restricted to stationary purchases on the e-commerce platform. Each student was allocated a R50 voucher as shown in Figure 1, which they were required to combine with group members to buy stationeries online. The product list was drawn up based on the charity center needs. This center offers ICT training to community members in Cape Town. This list was given to group members to select products they wanted to buy online based on their group budget. Students in the same group could share and combine vouchers to purchase more products as a group.

The project was divided into four phases as shown in Table 1. Activities in the project included engaging in digital practices, finding, evaluating and communicating online information, and collaboration. Students had to engage to search for products, evaluate product information on the e-business website and eventually buy the products using the supplied vouchers. It was envisaged that the process of buying products online would demonstrate the importance of analyzing and evaluating information such as prices, product description and delivery details before committing to an online transaction. Due to time constraints (Information systems is a semester course), the lecturer identified the supplier, and made sure that the products that students were going to buy, were available on stock. This was also a way of ensuring that the delivery would not be delayed due to unavailability of stock. Allocating shopping vouchers from a single e-commerce business unfortunately, limited students to buy from only one business.

Dividing students in groups encouraged teamwork, online communication (including transfer of files and vouchers between students) and collaboration. Students were also expected to apply problem-solving skills. They had to budget, depending on the available funds and the products they needed to buy. Students had to apply security and privacy concepts, for instance they needed to have the ability to distinguish the e-commerce suppliers' e-mails from possible fraudulent ones; and they had to ensure that their voucher numbers were kept confidential at all times. At the end of each deliverable, the lecturer allowed students to reflect on the process and provided feedback, and this was a way of scaffolding student learning of the e-business processes. The products bought as part of this project were donated for charity to a community project.

Digital literacies Integration into Curriculum

As noted above, the information system course content presupposes the use of ICTs for teaching and learning of the course. Hence, it is important to note that the focus here is broader than ICT integration into the curriculum.

Firstly, some tension between how learning technologists and literacies researchers use the term 'digital literacies' has been noted in the literature. Learning technologists tend to associate digital literacies with the affordances offered by technology tools while the literacy researches tend to focus on 'textual practices' in digital contexts (Lea, 2013). It is in this context that Martin (2006) emphasizes that the literacies (except ICT literacy) that form part of digital literacies, originated prior to the digital era but are now drawing on digital tools and environments, which have unfortunately, had a transforming impact on how people perceive digital literacies. Due to this, different terminologies have been used in literature for the digital literacy concept, such as information literacy; computer literacy – information technology or electronic information literacy; library literacy; media literacy; network literacy or internet literacy or hyper-literacy; digital literacy or digital information literacy (Bawden, 2001:219). Lea (2013), a New literacy studies researcher, attempts to reclaim the term 'literacies' to mean 'textual practices', and digital literacies to describe ways of enacting these textual practices in digital contexts. She critics the Joint Information Systems Committee's (JISC, 2011) definition of digital literacy – "those capabilities which fit an individual for living, learning and working in a digital society" (p.110). She is concerned that this definition tends to disassociate literacy from texts, and the "specific and contextual processes of meaning- and knowledge-making practices" (Lea, 2013:110). She proposes that digital literacies focus on the enactment of textual practices in digital context, and specifically on meaning-making and knowledge production. This debate about the uneasy relationships between learning technologies and literacy research is on- going (see Gourlay, Hamilton & Lea, 2014).

Secondly, there has been a growing tendency in the literature, of using the term 'literacy' synonymously with 'competence' or 'ability. For instance, Goodfellow (2011) explains that, in higher education practice, "the term [digital literacy] tends to refer to the competences of individual [students], rather than to literacy practices that serve the interests of specific communities, such as the IT, library, study skills and academic disciplinary communities" (2011, 139-140). In his earlier work, Martin emphasized that "digital literacy involves the successful usage of digital competences within life situations" (Martin 2008:169). Martin asserts that "digital competence is a requirement for and precursor of digital literacy, but it cannot be described as digital literacy" (Martin 2008:167). He emphasized that "digital literacy involves the successful usage of digital competence within life situations" (Martin 2008:169). That is, "the concept of literacy is only meaningful in terms of its social context and that to be literate is to have access to cultural, economic and political structures of society" (Martin, 2008:156). This notion of literacy being a situated practice resonates with the work of Scribner and Cole (1981), and it has been further developed by literacies researchers such as Street who contest for an ideological model of literacy - It posits that literacy is a social practice, not simply a technical and neutral skill as opposed to an autonomous model of literacy - it works from the assumption that literacy in itself - autonomously; literacy will have effect on other social and cognitive practices (2001:7). That is, literacy varies from one social group to another, and what is recognized as literacy depends on the context of a particular group. New literacies studies researchers such as Lankshear and Knobel (2006; 2008) and Lea (2013) have also pursued this trend the social practices trend, hence, they view of digital literacies being viewed as digital literacy practices that are relevant to a particular social and cultural context. For instance, Lankshear and Knobel describe literacy as a "family of practices—literacies—that include [...] socially evolved and patterned activities" (2008:256). Therefore, in this same context, Lankshear and Knobel recommend that digital literacy be perceived as a framework for integrating various other literacies and skill-sets, that (1) does not need to encompass them all; (2) cannot be reduced to a finite number of linear stages; and (3) cannot be acquired in the same way by all people or, by one person over all his or her lifetime (2008:28).

In their attempt to accommodate both schools of thought, in the LLiDA project, Beetham et al. focus on digital literacy practices:

We use the term '(underpinning) practices' in the hope of side-stepping some of the debates about definition and philosophy that beset literacies research, and in particular the 'paradigm contest' between cognitive and socially situated accounts of learning. Our focus in the study is on the pragmatic challenges that face learners and the institutions and educators that seek to support their development in practice as more capable human beings (2009: 8-9).

In this vain, Beetham, McGill & Littlejohn (2009) differentiate literacy from skill or competence. They assert that 'literacy' involves:

- A foundational knowledge or capability, such as reading, writing or numeracy, on which more specific skills depend
- **A Cultural Entitlement:** A practice without which a learner is impoverished in relation to culturally valued knowledge
- **Communication:** Expressing how an individual relates to culturally significant communications in a variety of media
- **The Need for Practice:** Acquired through continued development and refinement in different contexts, rather than once-and-for-all mastery
- **A Socially and Culturally Situated Practice:** Often highly dependent on the context in which it is carried out
- **Self-Transformation:** Literacies (and their lack) have a lifelong, life wide impact. (2009:9)

Implications for the curriculum are that when students are being prepared for the workplace, the integration of digital literacies into the curriculum has to be authentic. Beetham, McGill & Littlejohn (2009) suggest that when learners are being developed for situated practices, there is need for:

- Providing authentic contexts for practice, including digitally-mediated contexts.
- Individual scaffolding and support.
- Making explicit community practices of meaning-making.
- Anticipating and helping learners manage conflict between different practice contexts.
- Recognizing and helping learners integrate their prior conceptions and practices (2009:16).

Drawing on Vygotsky's (1978) socio-cultural theory of learning and development, situated learning, integrating digital literacies into the curriculum provides potential for the students 'opportunity to learn the socio-cultural practices which they can later develop and a better opportunity for them to 'imitate' what they have learnt in their course, in the workplace. For Vygotsky, learning is situated, and development entails the development of human higher cognitive functions which comes about as a results of mediated interactions of a learner (student) with a more knowledgeable peer (or lecturer) in the learner's Zone of Proximal Development (ZPD). Therefore, in an educational context, learning activities need to be designed for students with varied levels of preparedness, where these students need to be supported in their individual zones of proximal development. Vygotsky describes the ZPD as: The distance between the actual developmental level as determined by independent problem solving and the level of potential development as determined through problem solving under adult guidance or in collaboration with more capable peers (1978:86).

That is, the gap between what students can confidently do on their own and what they have yet to learn. Bridging this gap requires a more knowledgeable other, which could be a lecturer, peers or tools (ICT in this case), to mediate learning. Once the student is confident enough to independently solve a problem or work on a given task, he or she can 'imitate' that in a similar context. For Vygotsky, "a person can only imitate that which is within her developmental level" (1978:89). In this context, "imitation presupposes some understanding of the structural relations in a problem that is being solved (Vygotsky, 1987). In the context of this case, it is envisaged that integrating digital literacies into the curriculum of information systems course would embed the necessary digital and information literacies and skills to prepare students for the workplace particularly-business environments. The socio-cultural learning theory guided the pedagogy in this case. That is, learning activities were used learning activities were therefore need to be supported in their individual zones of proximal development.

In the context of this case study, the authentic learning pedagogic approach was used to introduce the students to the socio-cultural environment of e-business. The DIL framework developed by the Joint Information Systems Committee (JISC) in collaboration with the Open University, UK was used as a guiding framework for integrating ICT and digital literacies into the curriculum.

Application of the Digital and Information Literacy Framework (DIL)

Numerous frameworks for ICT in education have been developed by various stakeholders (such as HEIs, UNESCO, etc.) across countries. Reporting on the ten ICT frameworks that Rosado and Bélisle (2006) reviewed, they noted that although the frameworks fostered "updating, upgrading and enriching everyone to cope with the new demands of an information/knowledge society" (p.26), they failed to provide a generalized pedagogical approach for integrating digital literacies into school and university curricula (p.11). Furthermore, subsequent to the fuzziness of digital literacy conceptualization, a handful of digital literacy frameworks have been developed (See: Beetham 2009 and 2016) Other frameworks include Learner experiences of the e-learning' programme to develop an e-learner model that is focused on a student development approach (Sharpe and Beetham 2010). This has been criticized of being techno-centric.

Following from the above discussion and various institutional practices, there are a variety of models of students' digital literacies (Belshaw, 2011; Hinrichsen & Coombs, 2014). Belshaw (2011) talks of eight qualities of digital literacy; a digital literate student has to be cultural, cognitive, constructive, communicative, confident, creative, critical and civic. Hinrichsen and Coombs, (2014) outline five resources of critical digital literacy, such as, decoding, meaning making, using digital texts, analyzing digital text, and managing reputation and building digital identity (persona).

The Learning Literacies for the Digital Age (LLiDA) project report mapped out components of digital literacy that were integrated into curriculum practices of HEIs studies, such as:

- **Learning to Learn, Metacognition:** For which the competences include reflection, strategic planning, self-evaluation, self-analysis organization (time, etc.).
- **Academic Practice, Study Skills:** For which the competences include comprehension, reading/apprehension, organization (knowledge), synthesis, argumentation, problem solving, research skills, academic writing, specific subject discipline skills as appropriate.
- **Information Literacy:** For which the competences include information identification, accession, organization, evaluation, interpretation analysis synthesis, application.
- **Communication and Collaboration Skills:** For which the competences include teamwork, networking, speaking and listening skills.
- **Media Literacy (also 'Visual' and 'Audio' and 'Video' Literacies):** For which the competences include critical reading, creative production.
- **ICT/Digital/Computer Literacy:** For which the competences include keyboard skills, ability to use: capture technologies, analysis tools, presentation tools, general navigation/user interface skills, adaptivity, and agility confidence/exploration.
- **Employability:** For which the competences include (and are associated with graduate attributes) self-regulation, team working, problem solving, and business and customer awareness innovation/enterprise.
- **Citizenship:** For which the competences include (and are associated with graduate attributes) participation and engagement, ethical behavior, political and social, managing identity and reputation.

According to the JISC (2014) model, digital literacies entail the seven elements:

- **Information Literacy:** Find, interpret, evaluate, manage and share information.
- **Media Literacy:** Critically read, and creatively produce academic and professional communications in a range of media.

- **Communication and Collaboration:** Participate in digital networks for learning and research.
- **Digital Scholarship:** Participate in emerging academic, professional and research practices that depend on digital systems.
- **Learning Skills:** Study and learn effectively in technology-rich environments, formal and informal
- **ICT Literacy:** Adopt, adapt and use digital devices, applications and services.
- **Career and Identity Management:** Manage digital reputation and online identity.

Because the particular course is about digital and information practices, the DIL framework was found to be most suitable one. The strength of this framework is that it can be used as a guide for the development of authentic tasks for different discipline subjects. The framework tested and evaluated in multiple departments at the Open University. The researchers used the JISC - Open University Learning Design Initiative (JISC-OULDI) project digital facilitation cards as a guide to integrating digital literacies into the curriculum.

The focus of the research was equipping students with digital literacy skills and supporting student digital literacy practices in an e-business environment in an attempt to develop the students' identities.

The authentic task was designed in the form of a project that was intended to expose students to e-business systems. The JISC-OULDI digital facilitation cards were used for designing the student digital literacy activities that were integrated into the authentic task.

Drawing on the Open University's definition of digital literacy, the DIL Framework presents digital literacy in the form of five competence areas;

1. Understand and engage in digital practices.
2. Find information.
3. Critically evaluate information, online interactions and online tools.
4. Manage and communicate information.
5. Collaborate and share digital content.

Competences for each of the areas identified above have been divided into levels and stages, as follows

1. Access / level 0.
2. Foundation digital practice stage / level 1.
3. Interactive and co-operative digital practice stage / level 2.
4. Personalized and collaborative digital practice stage / level 3.

5. Professional and digital identity stage / masters.

The researchers used the JISC (2011-2013) digital information literacy (DIL) framework, which has been widely used in the JISC projects and Open University, UK. In a 2008-2009 JISC project: digital and information literacies were embedded in the tasks for the various phases of the project. The next section discusses how the DIL framework was used to design the project and how the competence areas were used to develop student attributes.

SOLUTIONS AND RECOMMENDATIONS

To elicit student digital literacy development, the study followed a mixed methods approach that used a closed-ended questionnaire and a semi-structured interview. A mixed methods approach (Creswell & Clark, 2007).

A purposive sample of 36 students who were enrolled in the IS responded to the questionnaire and 12 of them participated in the focus groups. The focus groups were used for completeness, and particularly to elicit students' experiences. Likert scale questions, ranging from strongly disagree to strongly agree, were used in the questionnaire. Participation in the study was voluntary, and students signed informed consent. The researchers also sought ethical clearance from the ethics committees, and permission to work with students, from Student Affairs Unit. Anonymity has also been ensured. Quantitative data from the questionnaires was analyzed in Excel. Interviews were taped, transcribed and analyzed, and verbatim transcripts were used for reporting the qualitative data from the interviews.

The following is a summary of the competences acquired from this authentic project.

Skills Level

Twenty respondents agreed that the e-business project helped them improve their skills of using internet browsers. Most respondents (24) could successfully follow instructions for accessing and using digital tools and websites. Twenty-four respondents agreed that the project improved their ability to critically evaluate information. The results are depicted in Figure 2.

Practices Level

Thirty-two respondents agreed that they independently navigated the supplier's website. The 32 respondents agreed that they searched the database and identified the products (in the shopping list) from the supplier's website. The lecturer gave students a list of products to shop for. Thirty-three of the respondents agreed that they successfully established contact with the supplier. The supplier would send account confirmation or validation e-mails to the students (clients), and other e-mails about the products being purchased by the students, and delivery arrangements. Thirty-three agreed that they could easily identify the supplier's e-mails from other e-mails that could possibly be fraudulent. A student attested to this in the focus group:

In terms of maybe, relating to a theory chapter on security, I learnt how to detect phishing, ... at the end of the e-mail from [the supplier], there is a privacy seal, which means that this is an accredited e-mail. So, that way you learn how to detect whether you are being floated or not, or if it's genuine or not. So that's a skill I picked along the way. [Student 1]

Thirty-one participants indicated that they successfully shared a range of formats of information, such as formats of scanned documents and other online content, among the group members.

They could further claim their vouchers and pay for their orders on the supplier's site. Twenty-seven of them indicated that they could also identify other payment methods that could be used on the supplier's site. Twenty-five respondents agreed that in their groups, they were able to use various online tools to discuss ideas relating to the project. Seven respondents were not sure whether they could use such tools in their group. Twenty-four of them agreed that scanning and uploading project documents, such as invoices and receipts onto the Vula site also improved their ability to share documents online. Twenty-eight of the respondents also indicated that they were successfully able to transfer their voucher credit to the buyer in the group.

Academic Literacy

When students were asked how participating in a group project impacted on their collaboration skills, group leaders shared that they acquired skills such as group management skills (tolerance, flexibility, accommodating), leadership skills (set time-frames, organize group discussion) and coordination skills (coordinating online sharing of documents and voucher credit transfers). Members acquired skills such as planning (the purchasing activities and process), respecting other team members' opinions, collective decision-making skills, task management as they assigned roles in the groups, budgeting with respect to their voucher credit and the prices of the products they had to buy. Also in terms of what products to purchase, one of the students reported the following.

You had to weigh the pros and cons of buying only this, and buying only that. So, as a group, you had to think tactically in terms of what to buy. No, let's buy this, I don't think many people will buy that, let's go for this and that. [Student 2]

We had to exhaust our funds. We wanted to spend as much of our budget as we possibly could....the main priority was to spend as much as we could, and to get as many products as we could. [Student 3]

About 80% of the students hence agreed that engaging in this task has helped them change their digital practices online as shown in Figure 3.

Personal Attribute or Identity Level

Students indicated at the end of the project they could confidently purchase, and manage an electronic transaction. Students also commented that they understood referral programmes as a result of completing the project tasks.

[The supplier], there's a referral programme where, if you refer someone and then that person can, like, that person is a buying customer. Like, you can, ja, you get additional vouchers on top of that. So, what the supplier did was, they have certain quarters where they release all the extra voucher to the friends you refer to the website.

The above results demonstrated that students have acquired the necessary skills as a result of embedding digital literacy in the curriculum. Students particularly from previously disadvantaged background benefitted from this project because its authentic nature and relevance of the activities completed.

FUTURE RESEARCH DIRECTIONS

The chapter explains how students were supported in acquiring the digital and academic literacy skills within the Information systems discipline. Insights on how ICT can be used as a tool to facilitate integration of digital literacies were highlighted. The DIL framework was used as a lens through which to reflect on inclusive curriculum design and integration of digital literacies.

This chapter is a first step towards shaping an undergraduate curriculum by using authentic learning approach using the DIL framework. The chapter demonstrated how using the DIL framework, which may serve as basis for future research direction in the ever-changing digital environment, developed student skills. Similar frameworks can be used to study how digital literacies can be integrated into curriculum for inclusive education. Future studies can use additional competence in the DIL framework for curriculum design to enhance student's digital practices and authentic learning in higher education.

CONCLUSION

The chapter explains how students were supported in acquiring the digital and academic literacy skills within the Information systems discipline. Insights on how ICT could be used as a tool to facilitate inclusive curriculum design were highlighted. There is a potential in using information and communication technologies to enhance educational inclusiveness of first year higher education students who hail from predominantly less privileged socio-economic backgrounds.

The DIL framework was used as a lens through which to digital literacies could be integrated into the curriculum. The project task presented an opportunity for students to engage and learn about basic terminologies that were challenging to explain the theoretical aspects of the course.

This chapter illustrates a potential innovative pedagogical approach to addressing inequalities as a means of alleviating marginalization of students from disadvantaged societies. The research component of this intervention focused on whether students learnt the intended skills over with the period of the project. Further research could be conducted to determine other means through which these students learn digital literacy skills.

Using the digital information literacy framework (DIL) in particular for undergraduate courses has demonstrated how embedding the relevant projects in the curriculum can develop students' attributes.

REFERENCES

Bawden, D. (2001). Information and digital literacies: A review of concepts. *The Journal of Documentation*, *57*(2), 218–259. doi:10.1108/EUM0000000007083

Beetham, H., McGill, L., & Littlejohn, A. (2009). *Thriving in the 21st century: Learning Literacies for the Digital Age* (LLiDA project). Retrieved May 20, 2015 from http://www.caledonianacademy.net/spaces/LLiDA/uploads/Main/LLiDAreportJune09.pdf

Bélisle, C. (2006). Literacy and the digital knowledge revolution. In A. Martin & D. Madigan (Eds.), *Digital literacies for learning* (pp. 51–67). London: Facet Publishing.

Belshaw, D. (2011). *What is Digital Literacy? A Pragmatic Investigation. Doctorate in Education.* Durham, NC: University of Durham.

Bennet, L. (2014). Learning from the early adopters: Developing the digital practitioner. *Research in Learning Technology, 22*(21453), 1–13.

Bharuthram, S., & Kies, C. (2013). Introducing e-learning in a South African Higher Education Institution: Challenges arising from an intervention and possible responses. *British Journal of Educational Technology: Journal of the Council for Educational Technology, 44*(3), 410–420. doi:10.1111/j.1467-8535.2012.01307.x

Bozalek, V., Gachago, D., Alexander, L., Watters, K., Wood, D., Ivala, E., & Herrington, J. (2013). The use of emerging technologies for authentic learning: A South African study in higher education. *British Journal of Educational Technology: Journal of the Council for Educational Technology, 44*(4), 629–638. doi:10.1111/bjet.12046

Brown, C., & Czerniewicz, L. (2010). Debunking the digital native: Beyond digital apartheid, towards digital democracy. *Journal of Computer Assisted Learning, 26*(5), 357–369. doi:10.1111/j.1365-2729.2010.00369.x

Chase, Z., & Laufenberg, D. (2011). Embracing the squishiness of digital literacy. *Journal of Adolescent & Adult Literacy, 54*(7), 535–537. doi:10.1598/JAAL.54.7.7

Cloete, N. (2014). The South African higher education system: Performance and policy. *Studies in Higher Education, 39*(8), 1355–1368. doi:10.1080/03075079.2014.949533

Creswell, J. W., & Clark, V. L. P. (2007). *Designing and conducting mixed methods research.* Retrieved May 25, 2015, from http://onlinelibrary.wiley.com

Czerniewicz, L., & Brown, C. (2013). The habitus of digital strangers in higher education. *British Journal of Educational Technology: Journal of the Council for Educational Technology, 44*(1), 44–53. doi:10.1111/j.1467-8535.2012.01281.x

Fisher, G., & Scott, I. (2011). *The role of Higher Education in Closing the Skills Gap in South Africa.Closing the Skills and Technology Gap in South Africa: Background Paper 3.* World Bank Human Development Group, Africa Region.

Goodfellow, R. (2011). Literacy, literacies and the digital in higher education. *Teaching in Higher Education, 16*(1), 131–144. doi:10.1080/13562517.2011.544125

Gourlay, L., Hamilton, M., & Lea, M. R. (2014). Textual practices in the new media digital landscape: Messing with digital literacies. *Research in Learning Technology,* 21.

Haywood, J., Haywood, D., & Macleod, H. (2006). Graduate e-literacies and employability. *Digital Literacies for Learning*, 123.

Hinrichsen, J., & Coombs, A. (2014). The five resources of critical digital literacy: A framework for curriculum integration. *Research in Learning Technology*, 21.

JISC. (2014). *Joined-up approaches to digital literacy development to support student achievement and employability: Developing digital literacies programme (2011-2013).* Retrieved May 21,2015 from https://www.jisc.ac.uk/guides/developing-students-digital-literacy

Joint Information Systems Committee. (2013). *Library Digital Facilitation Cards.* Retrieved June 22, 2015 from http://jiscdesignstudio.pbworks.com/w/file/63019416/Library_Digital_Faciltation_Cards_New_v23.pdf

Kajee, L., & Balfour, R. (2011). Students access to digital literacy at a South African university: Privilege and marginalisation. *Southern African Linguistics and Applied Language Studies*, *29*(2), 187–196. doi:10.2989/16073614.2011.633365

Knobel, M., & Lankshear, C. (2008). Digital literacy and participation in online social networking spaces. *Digital literacies: Concepts, policies and practices*, 249-278.

Lankshear, C., & Knobel, M. (2006). *New Literacies: Everyday Practices and Classroom Learning Maidenhead.* Open University Press.

Lankshear, C., & Knobel, M. (2008). Digital Literacy and Participation in Online Social Networking Spaces. In C. Lankshear & M. Knobel (Eds.), *Digital Literacies: Concepts, Policies and Practices. New York: Peter Lang.*

Lankshear, C., & Knobel, M. (2008). *Digital literacies: Concepts, policies and practices* (Vol. 30). Peter Lang.

Lau, J. (2006). *The impact of information competencies on socio-economic development in Southern Hemisphere economies.* Digital Literacies for Learning.

Lea, M. R. (2013). Reclaiming literacies: Competing textual practices in a digital higher education. *Teaching in Higher Education*, *18*(1), 106–118. doi:10.1080/13562517.2012.756465

Martin, A., & Madigan, D. (2006). *Digital literacies for learning*. Library Association Publishing.

Oyedemi, T. D. (2012). Digital inequalities and implications for social inequalities: A study of Internet penetration amongst university students in South Africa. *Telematics and Informatics*, *29*(3), 302–313. doi:10.1016/j.tele.2011.12.001

Prensky, M. (2001). Digital natives, digital immigrants part 1. *On the horizon*, *9*(5), 1–6. doi:10.1108/10748120110424816

Saunders, M. N. (2011). *Research methods for business students (5th ed.)*. Pearson Education India.

Sharpe, R., & Beetham, H. (2010). Understanding students' uses of technology for learning: towards creative appropriation. *Rethinking learning for the digital age: how learners shape their experiences*, 85-99.

Street, B. V. (2001). Ethnographic perspectives on literacy. In Literacy and Development: Ethnographic perspectives. London: Routledge.

Venkatesh, V., Croteau, A.-M., & Rabah, J. (2014). Perceptions of Effectiveness of Instructional Uses of Technology in Higher Education in an Era of Web 2.0. In *System Sciences (HICSS), 2014 47th Hawaii International Conference on* (pp. 110–119). Retrieved from ieeexplore.ieee.org

Vygotsky, L. S. (1978). *Mind in Society: The development of higher psychological processes*. London: Harvard University Press.

KEY TERMS AND DEFINITIONS

Digital Literacy: Digital literacy encompasses a range of practices, including computer literacy, information literacy, media literacy, communication literacy, visual literacy and technology literacy.

E-Commerce: The Buying and selling of goods online.

E-Business: Conduct of business process online or the internet.

JISC-OULDI: Digital facilitation cards were used for designing the student digital literacy activities that were integrated into the authentic task.

Information and Communication Technology: Umbrella term used for information and communication devices such computers, used to share information.

APPENDIX 1

Figure 1. Example of a voucher purchased for students

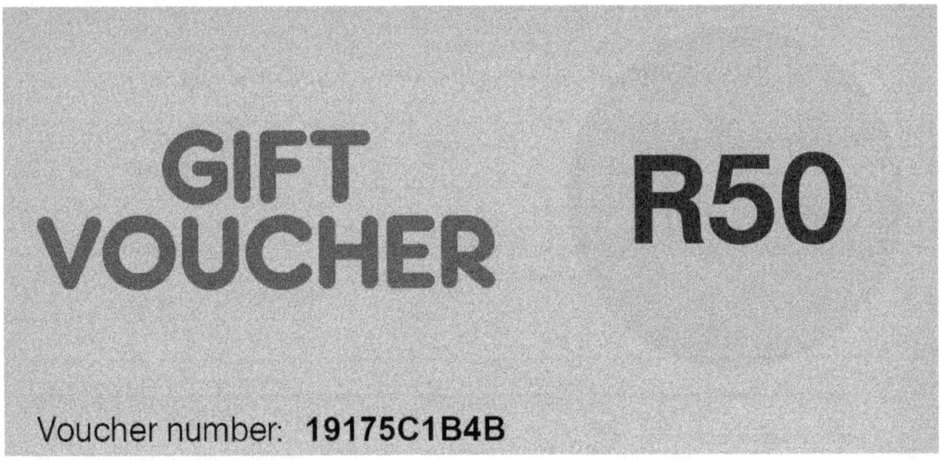

APPENDIX 2

The student task was divided into project phases.

Table 1. Deliverables for the project phase

Phase	Short Description	Assessment of Artefact
1: Signing up on the e-business website	Create an account on e-commerce website, demonstration of importance of keeping account information confidential	Upload screenshot of proof of registration or account/ profile created, onto the LMS course site
2:Evaluation of products and Purchasing	Evaluate product prices, add items to shopping cart, transfer of shopping vouchers, confirmation of purchased items and invoices	Claim individual vouchers Submit invoice onto LMS course site
3: Delivery	Verification of delivery address, ensure products are delivered to the university address	Take a picture of delivery receipt and submit onto the LMS course site
4: Presentation	Develop communication and presentation skills of students	Presentation of the project and products in class

APPENDIX 3

The project made use of these concepts from the framework to integrate the literacies in the project task.

Table 2. The Digital Information literacy framework and applied concepts

DIL Competence Area	Description	Example from the Learning Activity
Developing Internet navigation and search skills		
Finding information	Demonstrate ability to plan and adapt a search [Level 1]	Navigating the Internet and creating an account on the e-business website
Familiarise student with the course LMS tools and		
Finding information	Successfully carry out instructions for using a range of formats of information [Level 2]	Capturing screenshots of registration confirmation and invoices, and taking pictures of delivered products and uploading onto the course site; invoice
Understanding and engaging in digital practices	Successful use of the LMS (course site) [Level 1] Demonstrate ability to produce subject-related artefacts using digital tools and resources [Level 2]	
Collaborating and sharing digital content	Successfully follow instructions for sharing digital content with multiple recipients using the LMS [Level 1] Successfully negotiate roles and carry out own role within a task of a teacher- or student led online activity [2]	Transferring vouchers among group members Students claimed their purchasing vouchers from the voucher system managed by the lecturer Students also could either meet physically or collaborate online for the group work
Course-related aspects: 1) Familiarizing students with security and privacy practices		
Understanding and engaging in digital practices	Demonstrate ability to manage shopping account Successfully verify e-mail communications with various stakeholders	Students had to manage their shopping accounts and verify e-mail communications with the supplier
Develop information		
Manage and communicate information		Students managed the different formats of information, designed group presentation and presented in class

APPENDIX 4

Figure 2. Skills level

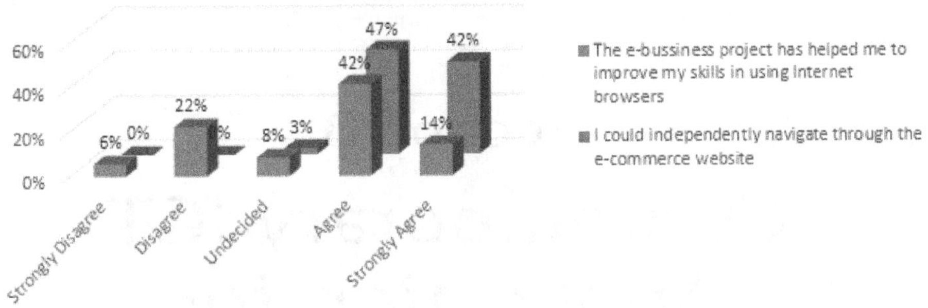

APPENDIX 5

Figure 3. Students digital practices

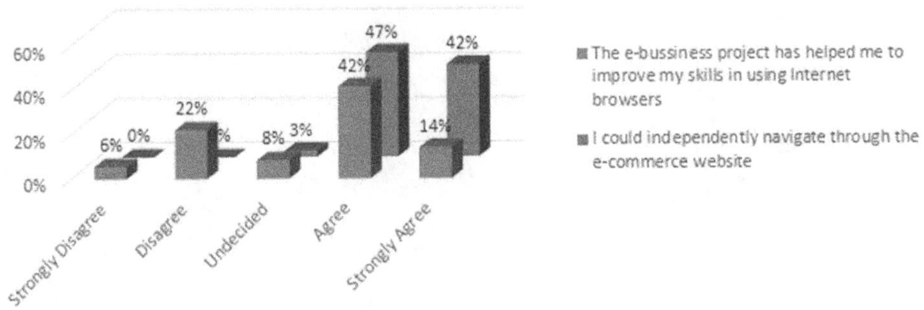

Section 3

Contemporary ICT Application for Development

Chapter 8
A Cloud Computing Adoption Framework for Developing Countries

Anh D. Ta
Advanced Technology Consulting, LLC, USA

ABSTRACT

This chapter explores Cloud computing adoption strategies for small and medium enterprises (SMEs) and microenterprise that can enhance the economic growth of local industries in developing countries. This focus is based on the recognitions that (1) economic growths of local industries reduce poverty through the creation of new jobs for poor people to earn a living wage and support their families, and (2) there is a lack of research on Cloud computing adoption strategy that addresses the unique constraints of small and medium enterprises (SMEs) in developing countries.

INTRODUCTION

This chapter presents a cloud computing adoption framework for developing countries that small and medium enterprises (SMEs) and microenterprises can use to enhance the economic growth of local industries through the creation of new jobs with living wages. The framework incorporates proven strategies with the potential to quickly grow local industries, specifically adopting advanced technologies to narrow the productivity and research gaps. The framework has two dimensions. The first dimension represents the requirements across the different economic levels. The second dimension represents the cloud computing characteristics to mitigate

DOI: 10.4018/978-1-5225-2565-3.ch008

Table 1. High-level tradeoff for industry type

Industry Type	Advantage	Disadvantage
Manufacturing	• The employment of a large population. • It contributes revenue to the government via taxes.	• The required financial investment is large. • There is the potential of pollution from improperly managed manufacturing process. • There may be long-term environmental effects from the construction of factories.
Natural Resources Exploration	• The employment of a large population. • It contributes revenue to the government via taxes.	• The required financial investment is large. • There is the potential of pollution from improperly managed manufacturing process. • There may be long-term environmental effects from the excavation process.
Tourism	• The employment of some the population as service employees. • Tourism-related businesses will be created along with new jobs.	• The available jobs are short-term, non-skill positions.
Information Technology	• The population is employed and trained for high demanding jobs. • The jobs are long-term because of project global demand for IT professionals.	• The required financial investment can be large depending on type of project/initiative. • Some period of training is required to properly support IT jobs.

adoption risks in developing countries. An example set of cloud computing technology products will be presented to illustrate the potential benefits of the framework.

BACKGROUND

Economic Growth Strategy for Developing Countries

Economic growth in developing countries is the most important means of raising income and reducing poverty through the creation of new jobs for poor people to earn living wages and support their families; the result is a more stable future for the country. However, the available options of adoption of a certain industry typically involved complex tradeoffs such as environmental and social side effects. Table 1 lists the example advantages and disadvantages for industry types.

Furthermore, "[m]any developing countries face particular challenges that make it difficult for them to stimulate and sustain economic growth. These challenges include weak institutions, high unemployment, poor infrastructure, a lack of access to financial services and unsuitable laws and regulations. Options for developing countries are limited." (Greening, 2015) Besides the limited options, the window of opportunity for economic growth is narrowing fast. Research results have shown (Comin & Hobijn, 2012; Nobel, 2012) that "...the rate at which countries adopted new tools hundreds of years ago strongly affects whether they are rich or poor today." More importantly, the same research (Comin & Hobijn, 2010; Nobel, 2012) also identifies a correlation that technology adoption has long-term effects on the economy, where "[t]he longer the lag in technology adoption for any given nation, the lower the per capita income."

While these research findings highlight formidable challenges to developing countries, other researchers (Manyika et al., 2013; Foster & Rosenzweig, 2010) have identified a risk mitigation approach to address these challenges where "[o]ne mechanism by which poorer countries can catch up with richer countries is through technological diffusion, the adoption by low-income countries of the advanced technologies produced in high-income countries." (Foster & Rosenzweig, 2010). This technology diffusion approach called *leapfrogging* is often used to "...narrow productivity and output gaps, and reduce research and development cost." (Hashim et al., 2014) In a holistic approach to technology diffusion, it is important to address both intensive and extensive margins of technology adoption, where *extensive margin* refers the time it takes for a country to adopt a technology and the *intensive margin* refers to the extent to which a technology is adopted by the population (Comin & Mestieri, 2010). These facets are important because "...while a new technology may reach a third-world country faster than ever before, it's not necessarily reaching the majority of the people in the country." (Nobel, 2012). For example, the lack of ownership or access to computers prevents the population in developing countries access to free software services such as webmail, websites, etc.

Assuming that Information Technology (IT) industry is the more preferable choice to adopt for developing countries versus traditional manufacturing industries, then, the questions become which components of the IT industry should a developing country adopt and what would the adoption strategy be? A recent McKinsey Global Institute report (Manyika et al., 2013) identifies twelve technologies that have the greatest potential to drive substantial economic impact and disruption by 2025, where three Internet-related technologies are mobile Internet, the Internet of Things (IoT), and cloud technology (or cloud computing). Of these, mobile Internet and cloud computing are most applicable for developing countries because of current technology adoption trends and leapfrogging opportunities. The McKinsey Global Institute report (Manyika et al., 2013) estimated that by 2025 there will be:

- 4.3 billion People still remaining to be connected to the Internet, potentially through mobile Internet.
- 2 billion global users of cloud-based email services such as Gmail, Yahoo, Hotmail, etc.
- The monthly cost of renting in the cloud will be three times cheaper versus owning a server.

- $1.7 trillion of the global Gross Domestic Product (GDP) will be related to the Internet.

Based on the global forecast of cloud computing, it is recommended that developing countries should consider adopting it for economic growth.

The next two sections provide an overview of virtualization technology and cloud computing to establish the terminologies and concepts that will be presented in the proposed framework.

Virtualization Technology

Virtualization is a software technology that abstracts the physical resources of computers or servers to support the concurrent execution of multiple operating systems and applications. This concept originated in the late 1960s and early 1970s when IBM introduced and evolved the time-sharing paradigm in mainframe computers. The shared usage of computer resources among a large group of users increases the efficiency of both the users and the expensive computer resources they share. (Oracle, 2013)

Today, virtualization technology is making a resurgence in the form of hypervisor (or Virtual Machine Monitor [VMM]). The hypervisor creates a virtual platform on the host physical computer, on top of which multiple guest operating systems (or Virtual Machines [VMs]) are executed and monitored. In this context, the operating system instances share hardware resources owned by the host physical computer.

There are two types of hypervisor:

- **Type 1 Bare-Metal:** The hypervisor is installed directly on the physical hardware without an operating system.
- **Type 2 Hosted:** The hypervisor is installed on top of an operating system similar to other installed software products.

A hypervisor serves two main functions. The first function is to provide the virtualization environment that abstracts resources on the physical hardware as virtual resource pools. The second function of a hypervisor is to manage the VMs, where a VM is an emulation of a physical server using resources from available virtual resource pools. This is the essence of server virtualization.

To scale a virtual environment, multiple hypervisors can be configured as a cluster. The hypervisor cluster offers high availability support for VMs by automatically load balancing the VMs' processing among clustered host machines (e.g., if a host machine is down then the hypervisor cluster automatically moves the VMs from the unavailable host to other available hosts). As the result, there is no need for a cluster configuration of VMs at the operating system level in a virtual environment. This feature reduces configuration complexity for servers. Another benefit of a VM is the flexibility of changing the allocated resources to reflect the true requirements of a server versus allocating a fix configuration; this is referred to as *right sizing* a server. These are example benefits of the reduced complexity in managing virtual servers as compared to physical servers.

The concept of desktop virtualization extends the virtualization technology to another usage level where user-facing desktops are hosted as VMs. Virtual Desktop Infrastructure (VDI) software provides and manages the virtual desktops using both the VDI Hypervisor and the VDI Provisioning components. VDI offers the benefits of centrally managing all desktop configurations via VMs and offering users various desktop configurations through different desktop pools (i.e., VDI Connection Broker). Users access the virtual desktop via an agent installed on the user's device and the agent communicates with the virtual desktop using the remote presentation protocol, presenting the virtual desktop as though it was actually running directly on the user's device.

Application virtualization is highly recommended in order to fully realize the benefits and flexibility of VDI. Application virtualization is a software technology that encapsulates application software from the underlying operating system on which it is executed. Specifically, application virtualization creates a container for the software that contains all information typically stored in an operating system or a file system component like a system registry. The application container provides a self-contained, portable environment that minimizes conflict with other installed software products. It helps VDI bring value to the enterprise by providing a simple mechanism to share an application among desktops without worrying about inconsistent desktop configurations.

With an overview of virtualization technology, the next section describes how virtualization technology is making everything possible for cloud computing.

Cloud Computing

The United States National Institute of Standard and Technology (NIST) defines *cloud computing* as "...a model for enabling ubiquitous, convenient, on-demand network access to a shared pool of configurable computing resources (e.g., networks, servers, storage, applications, and services) that can be rapidly provisioned and released with minimal management effort or service provider interaction." (Mell & Grance, 2011). In the context of cloud computing, the enterprise services that typically reside at on-premises, physical IT infrastructure are now hosted at a network accessible abstracted location (or "the cloud"), where the physical location and the underlying hardware are irrelevant. This shift in enterprise service model is analogous to the utility model where a homeowner switches from using in-house electric generator to using electricity from a local electric utility company for enhanced safety, scalability, and efficiency along with other benefits. One key financial benefit of the service model is the concept of paying only for what you use.

Based on a cloud location, we classify a cloud as:

- **Private Cloud:** Has the virtual environment hosted within an enterprise and all services are available only via the enterprise network.
- **Public Cloud:** Has the virtual environment hosted by a Cloud provider and all services are accessible from both the enterprise network and the Internet.
- **Hybrid Cloud:** Has the virtual environment spread between both local virtual environment and remote virtual environment from Cloud provider.

Based on a service that the cloud is offering, the service model includes a range of virtual infrastructure to support:

- **Software as a Service (SaaS):** A delivery method that provides users access to application remotely. Example of application categories are mail clients (e.g., Gmail, Hotmail, Yahoo Mail, Microsoft Outlook Web Access), office applications (e.g., Microsoft Office 365, Google Docs), storage (e.g., iCloud), etc. A SaaS provider performs the maintenance and support for the services while the users only have access to the account and the user's content.
- **Platform as a Service (PaaS):** A delivery method that provides users with server resources to develop or host applications. A PaaS provider hosts the required hardware and software on their infrastructure. As a result, users do not have to install in-house hardware and software to develop or run a new application. Examples of products offering PaaS are Web servers (e.g., SquareSpace), database (e.g., Microsoft SQL Azure, Amazon RDS), etc.
- **Infrastructure as a Service (IaaS):** A delivery method that provides users with servers and storage resources. An IaaS provider performs tasks such as system maintenance, backup and resiliency planning while the user is responsible for the application installed on the server.

The cloud computing service model provides enterprise with a strategy that can quickly and dynamically scale service capability while maximizing the limited budget and availability of trained personnel.

MAIN FOCUS OF THE CHAPTER

Issues, Controversies, Problems

While there are many documented cloud computing adoption strategy and lessons learned from developed countries, there is few cloud computing adoption strategy that addresses the unique constraints of small and medium enterprises (SMEs) and microenterprises in developing countries (Adam & Musah, 2014; Miliaras, 2012). A review of literature shows that "the migration process of SMEs to the cloud and how this impacts existing process and the sustainability of the business has is completely missing." (Adam & Musah, 2014). In fact, some cloud computing adoption strategies try to apply the same assumptions of developed countries to developing countries. Charles Kenny, a senior fellow at the Center for Global Development reminds us that "the internet and digital technologies offer huge potential but avoiding a one-fits-all approach and ensuring the right protection mechanisms is vital." (Kelly, 2013). Overall, there is a need for a cloud adoption strategy for developing countries that ensures both wide spread diffusion and substantial implementation successes (i.e., extensive and intensive margins).

This chapter focuses on the cloud computing adoption strategy for small and medium enterprises (SMEs) that can enhance the economic growth of local industries in developing countries and not the general adoption of cloud computing by the general population (e.g., adoption of cloud services hosted by companies outside of the developing countries, adoption of cloud services relating to social media). This limited focus is based on the recognition that economic growths of local industries reduce poverty through the creation of new jobs for poor people to earn living wages and support their families.

Risks to Cloud Computing Adoption in Developing Countries

In developing countries, there are many risks to the adoption of cloud computing involving both technical and social factors. An overview of each risk is provided below to establish the problem space constraints to be addressed:

- **Access to Reliable Electrical Power:** A fundamental risk to cloud computing is the availability of reliable electrical power. This risk is not unique to developing countries but it is more profound because of the uneven economic and infrastructure development across the countries. The United Nations Sustainable Energy for All initiative (SE4ALL, 2015) estimates that over 1.3 billion people—one in five globally—lack electricity to light their homes or conduct business. Of those, over 95% of them live in sub-Saharan Africa or developing Asia. The *Poor People's Energy Outlook 2013* (Practical Action, 2013) report published by Practical Action estimates that nearly half of all health facilities in India – serving an estimated 580 million people that lack electricity. Also, 30% of health center facilities in sub-Saharan Africa, serving 255 million people, are without electricity.

- **Access to Computer:** Another often cited risk to adopting cloud computing is the limited access to computers (e.g., laptop, desktop, smart mobile devices) that allow users to leverage cloud services. The main reason for the limited access to computers is cost. This risk is being addressed by many initiatives (e.g., Youth Learning by Dell[1]) that make computers, low-cost laptops or tablets available to the population. However, the access to computer without an IT infrastructure and program to enhance computer literacy will result in locally optimized benefits. Deb Bauer, director of Dell Giving substantiates this observation by saying that "...what we've learnt is that it isn't enough to simply provide the hardware, it's the quality of the wrap-around services – the teacher training, maintenance of technology, reliability of power, which provides the long-term benefits and this is one of the learnings we've been taking forward." (Kelly, 2013).
- **Computer Literacy:** There is a need for educational initiatives to enhance computer literacy so that individuals and businesses can improve productivity with automation via computer and cloud computing services. The degree of computer literacy varies from no experience to highly experienced. A recent Pew Research (Pew Research Center, 2015) study found that "Once online, internet users in emerging and developing nations have embraced socializing as their most preferred type of digital activity…Less common are commercial and career activities, such as searching or applying for a job, making or receiving payments, buying products and taking online classes." The goal of enhancing computer literacy should be on developing skills relevant to a professional career versus social media and information search; while these are valid usage scenarios, they do not directly contribute to the economy of developing countries.
- **Internet Connectivity:** The fact that a computer literate population has access to computer is of limited benefits if there is unreliable broadband Internet connectivity. The state of Internet connectivity infrastructure in developing countries is similar to the state of their physical infrastructure, which is inconsistent and unreliable. According to the World Internet Users and 2015 Population Stats survey (Internet World Stat, 2014), Africa (27.5%)

and Asia (34.8%) has the lowest Penetration (i.e., percentage of population having access to the Internet) as compared to the North America (86.9%) and Europe (70.4%). Part of the problem is poor network performance. "The total bandwidth available to shuttle data between African countries and the rest of the world in 2011 was less than 1 terabit per second. That's about one-seventieth of Europe's international bandwidth capacity. Making matters worse, the price for bandwidth in Africa is cripplingly high. While a university in Germany might pay about US $4000 per month for 1 gigabit per second of bandwidth, a school in Kenya can expect to pay $200 000 for the same service" (Cottrell, 2013). Not all networks are applicable to cloud computing; however, many low-income countries rely on mobile broadband networks that are not ideal for cloud computing because of their low-speed.

- **Security:** From the security perspective, there are two risks to cloud computing adoption that must be addressed. First, the enforcement of physical security for deployed computer equipment or valuable resources in remote population. The worst-case scenario is to have criminals stealing the computers or related equipment after deployment. Second, the laws in developing countries may not have provisions or amendments ensuring privacy and protection of data. In fact, The Information Economy Report 2013 (UNCTAD, 2013) warns "… that the global shift towards cloud computing, which allows users to store and access data remotely, brings a range of legal as well as technological and infrastructure challenges for poor countries". The report also stresses that "…while there is no imperative to develop specific laws or regulations on cloud computing, key reform areas include privacy, data protection, information security and cybercrime. Governments of developing countries should adopt and enforce appropriate laws and regulations" (UNCTAD, 2013). The International Telecommunication Union (ITU) has a stronger recommendation that "a patchwork of inconsistent and largely unenforceable national and regional regulations will neither harness the opportunities presented by cloud services nor secure an individuals' private rights and information" (ITU, 2012).

These risks are related and varied across economic levels of developing countries. Hence, the risks require a holistic approach to mitigate and maximize the available resources. The subsequent sections will present the strategy and framework for cloud computing adoption that mitigates the risks described.

SOLUTIONS AND RECOMMENDATIONS

Adoption Strategy

Before proposing a cloud computing adoption framework for developing countries, a strategy for addressing the risks of cloud computing in developing countries is presented. The review of cloud computing adoption literatures identifies the main requirement is for constant Internet connectivity. However, a closer analysis of workers' habits in most industries reveals that the degree of network connectivity varies with the scope of a specific task. For example, a worker who is developing a document or a software program would need connectivity to access resources on enterprise servers but not constant Internet connectivity to access resources external to the enterprise. There are some situations that a worker would need Internet access for services such as e-mail or information search but those accesses may or may not be related to the enterprise operational goals. Hence, the network connectivity requirement varies with the scope of each task for an economy level (i.e., National/ Regional, SME, and Microenterprise). In addition to the varying connectivity requirements, there is also a recognition that cloud computing adoption risks vary between developing country economic levels. These observations highlight the need for a customized solution to address the unique microcosm of each developing country economic level.

In the context of this chapter goal, the proposed cloud computing adoption framework for developing countries has two foci:

1. **Adoption by SMEs and Microenterprise:** The proposed framework will emphasize more on the adoption process for SMEs and microenterprise because research data shows that they contribute the most to economic growth in developing countries with creation of new jobs. A research on generating jobs in developing countries (Demirgüç-Kunt, 2011) shows that SMEs (1) contribute more to employment in low-income countries than higher-income countries, and (2) have the largest share of employment in developing economies. Similar to SMEs, a *microenterprise* business employing 10 people or less have similar proportional contribution as SMEs to the larger economy. The successes of microenterprise are fueling the popularity of *microloans* across the world. In fact, one venture capitalist states that funding low-income entrepreneurs is "one of the most important economic phenomena since the advent of capitalism" (Stanford, 2004). The challenge within the proposed framework is to identify a solution that maximizes the benefits of cloud computing adoption in the context of limited resources (e.g., financial, trained personnel, etc.).

2. **Maximize Available Components:** The proposed framework will identify adoption strategy for each level of the economy that maximizes the benefits of available cloud computing components while addressing constraints from adoption risks. Research results have shown that the acceptance and adoption of cloud computing by SMEs in developing countries are slow due to several risks ranging from technological, environmental and organizational constraints (Hashim et al., 2014; Yeboah-Boateng, & Essandoh, 2014; Kapurubandara & Lawson, 2006). In general, cloud computing requires constant electricity and reliable broadband connectivity or internet connectivity. Unfortunately, these are not characteristics of developing countries infrastructure. The challenge within the proposed framework is to identify a solution that maximizes the benefits of functionality from available cloud computing components to addresses either extensive margin (or scope) and intensive margin (or intensity) of cloud computing adoption in developing countries.

Adoption Framework

This section presents a cloud computing adoption framework for developing countries that small and medium enterprises (SMEs) and microenterprises can use to enhance the economic growth of local industries through the creation of new jobs with living wages. The framework incorporates proven strategies with the potential to quickly grow local industries, specifically adopting advanced technologies to narrow productivity and research gaps. The framework has two dimensions. The first dimension represents the range of requirements across the different economic levels. The second dimension represents the adoption strategy characteristics to mitigate cloud computing adoption risks in developing countries. An example set of cloud computing technology products will be presented to illustrate the potential benefits of the proposed framework.

Table 2 represents the proposed cloud computing adoption framework for developing countries as a matrix. The first dimension represents the Economic Levels as columns in the matrix. The second dimension represents the Adoption Characteristics as rows in the matrix. The characteristics are prioritized according to both their importance to successful cloud computing solutions and the solution flexibility with regards to number of available options. Both cloud type / services and power source characteristics are assigned a high adoption priority because they represent the core components of successful cloud computing solutions. The network connectivity and data storage characteristics are assigned a medium adoption priority because there are few options available for both local and regional levels (e.g., WiFi can be used for local connectivity and scheduled internet access is used to transmit

Table 2. Cloud computing adoption framework for developing countries

			Economic Level Dimension		
			National / Regional	Small and Medium Enterprise (SME)	Microenterprise
Adoption Characteristic Dimension	Adoption Priority	Adoption Objective			
	High	Cloud Type / Services			
	High	Power Source			
	Low	Client Device			
	Medium	Network Connectivity			
	Medium	Data Storage			

files or obtain software updates). The client device characteristic is assigned a low adoption priority because there are many low-cost options to address user access requirements. Overall, the matrix content will describe the recommended cloud computing adoption strategy for an economic level. Each strategy will be discussed below starting with the local or microenterprise level.

Microenterprise

As described, the successes of microenterprises contribute significantly to a developing country's economy. Microenterprises create new jobs and support the increase in computer literacy among the population. Furthermore, microenterprise can be targeted at remote population as part of the strategy of evenly distributing economic initiatives throughout a country.

In this context, the cloud computing adoption objective for microenterprise should be to establish businesses offering internet café like services in remote locations to expose the population to information technology and promote entrepreneurship along with academic pursuits (e.g., freelance document translation, online course). These remote locations may have some access to cloud services (e.g., e-mail, web sites) via smart phones. However, these internet accesses are not related or supported in the professional development of the population. Hence, the recommended virtual infrastructure for microenterprises is to establish a micro, private cloud that offers virtual desktop infrastructure (VDI) services to the population. This private, micro cloud can host locally selected, structured contents such as electronic books, movies, reference materials (encyclopedia), educational games, etc. This recommendation assumes that users who are new to computer and the internet only need certain structured content versus sophisticated users who can independently navigate among internet resources. This structured approach is not promoting censorship but ensuring that users have positive experiences. In fact, this selected, structured initial access to information via micro cloud in some cases can avoid problems from encountering content on the internet that conflicts with local religious teaching or social norms in developing counties.

With today's technological advancements for both hardware and software, a portable, private, micro cloud (See Figure 1) can be deployed using a simple configuration consisting of a network router with WiFi and laptop with (1) processor supporting virtualization, (2) adequate RAM memory and (3) storage. For example, a laptop with i7 processor, 8 GB of RAM, and 500 MB of disk storage and Oracle VirtualBox, an open source software, can support 10-20 small virtual machines or virtual desktops. The primary power source can be a solar powered generator like Goal Zero: Yeti 150 Solar Generator Kit. It should be noted that the laptop battery can last several hours and can be recharged using a solar powered generator when needed. This configuration minimizes the need for constant power source like the national grid. The micro cloud configuration should also include an uninterruptible power supply (UPS) to provide users with enough time to save their current processing or products, supporting a data loss prevention strategy. Since the micro cloud configuration has few components, it is portable and suitable for deployment to remote locations.

As for user access to virtual desktop, there are two options: (1) deploy LCD/LED display with Micro PC or (2) deploy Chromebook, tablet, and/or smartphone. Option 1 leverages the benefits of low power consumption by both the LCD/LED display and the Micro PC to provide user access to virtual desktops. Another benefit

Figure 1. Micro cloud configuration

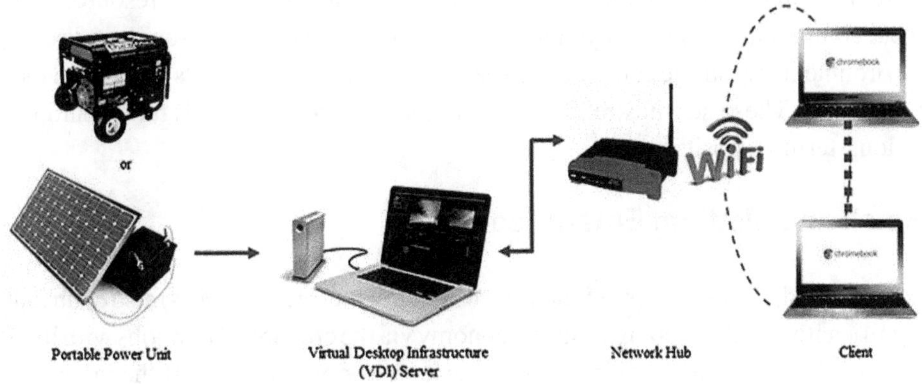

is the low cost of Micro PC. Since only a thin, Web browser client is required to access remote virtual desktops, it is possible to use old or existing desktop PCs to support user access requirements. However, the power consumption requirements of the old desktop PCs may not be the optional solution. Also, the cost of the LCD/LED display may be too expensive for developing countries. Option 2 may be a more cost effective option for developing countries based on the low price of each Chromebook and tablet along with the fact that these devices have both display and keyboard. In Option 2, users would access the virtual desktops via WiFi using Chromebooks. For both options, users should store their personal data on portable Universal Serial Bus (USB) flash drive. This approach to data storage mitigates the privacy and data security risks in developing countries until there are strong privacy and data security laws in place. Table 3 summarizes the cloud computing adoption strategy recommended for microenterprises.

The proposed micro cloud configuration for microenterprises has the potential for dual-use benefits in supporting the cloud computing adoption and United Nations' SE4All initiative such as providing lighting for medical operation, powering refrigerator to store medicines (WHO, 2015; Adair-Rohani, H. et al., 2013). Specifically, the unused computing and/or electrical power can be shared with local community and community institutions such as school, health center, etc. A possible scenario of dual-use of micro cloud configuration is to use the computing services during daytime for paying customers and unused resources for local school computer lab or local health facilities. Then in the evening, the computer services can be used to support adult education programs as part of a community outreach initiative. In addition, the unused electricity can support clean LED lighting that is much safer then kerosene lamps and candles (Grimshaw & Lewis, 2010). An alternative application of the unused electricity is to partially address the cooking needs of those who are still cooking on harmful and inefficient stoves (IEA, 2014). This dual-use energy access approach would be a more economical option than extending the national grid to the remote locations (Bahaj, 2016).

In addition to the benefits from sharing computing and electric resources with local communities, the deployed micro cloud configuration can provides a training environment for routine computer and network administration skills to selected local population. These administration skills are the high demand skills that would lead to long-term IT positions.

Small and Medium Enterprise (SME)

As described, the successes of Small and Medium Enterprises (SMEs) also contribute significantly to a developing country economy via the creation of new jobs with living wages. SMEs support local industries, create new jobs, and support the increase in

Table 3. Microenterprise cloud computing adoption strategy

		Economic Level Dimension	
		Microenterprise	**Example Configuration**
Adoption Characteristic Dimension	**Adoption Objective**	Offer local population Internet café with virtual desktop to access computing resources to narrow the digital divide and enhance digital literately.	
	Cloud Type / Services	• Private, micro cloud • Virtual machines • Virtual desktops	• A laptop with - i7 processor - 8 GB of RAM - 500 MB of disk storage - $1000 • Open source hypervisor - Oracle VirtualBox - Citrix XenServer
	Power Source	• Primary - Solar power with battery backup • Secondary - Mini/Micro Grid - UPS w/ 10 minutes backup time - Alternative energy sources	• Goal Zero: Yeti 150 Solar Generator Kit - $500
	Client Device	• LCD/LED display with Micro PC • Portable devices - Chromebook - Tablet - Smart phone	• Micro PC - ZOTAC ZBOX PI320 pico - Atom Z3735F 1.33 GHz - 2 GB RAM - 32 GB SSD - Intel HD Graphics - Windows 8.1 - $169 • Chromebook ($150-$200) • Android tablet ($30-$150)
	Network Connectivity	WiFi with Internet / broadband connectivity on an as needed basis.	• 4 ports network router with WiFi ($15-$70)
	Data Storage	Local external personal storage device such as USB flash drive.	• 16 GB USB flash drive ($5-$15)

computer literacy among the population. The ultimate goal is to support SMEs so that they can support local industries with export capacity to enhance the developing country's Gross Domestic Product (GDP) and possibly narrowing the trade gaps.

In this context, the cloud computing adoption objective for SMEs should support IT outsourcing so that SMEs can focus on the operational efficiency and growth of its core business. Hence, the recommended virtual infrastructure for SMEs is to use a hybrid cloud configuration where the private cloud consist of a SME's virtual infrastructure and the public cloud would involve services (i.e., SaaS, PaaS, IaaS, and VDI) from multiple cloud services providers. Specifically, the SMEs would host the content critical to their daily business operations in the private clouds and other static or long-term content in the public clouds. This recommendation is to mitigate the risks associated with unreliable Internet connection along with privacy and data security. Specifically, the hybrid cloud configuration allows SMEs to continue daily operation using the private cloud services when Internet connectivity is not available. Furthermore, the option of storing data in private cloud or pubic cloud allows SMEs to perform tradeoffs on the volume of data to store on cloud as a reflection of their data security risk aversion level. The proposed hybrid cloud configuration would require coordination with local government on public cloud initiative investments.

For SME's private cloud, they should perform tradeoffs between deploying physical servers or latest technology like hyper-converged server, a virtual computing infrastructure solution that seamlessly combines several data center services in an appliance form factor – a technology leapfrog opportunity. Specifically, the hyper-converged server is a preconfigured server and storage components to support virtual data center services out-of-the-box suitable for remote office/branch office (ROBO). This seamless integrated appliance accelerates the speed and agility of deploying virtualized workloads, reduces complexity, improves operational efficiency, and lowers costs. Also, it offers a scalable building-block approach allowing IT department to expand capacity by adding units. This granular scalability is one of the hallmarks of hyper-converged infrastructure (Lowe et al., 2016).

The primary power source for SMEs should be the national grid augmented with mini/micro grid to mitigate risks of unreliable power in developing countries' infrastructure. As for secondary power sources, SMEs should also use micro grid or alternative energy sources (e.g., solar powered generators, gas powered generators) when available to mitigate short duration power outages. Furthermore, SMEs should deploy UPS as part of the data loss prevention strategy to provide enough time for users to save current processing or work products.

As for end user devices, there are three options: (1) deploy LCD/LED display with desktop PC, (2) deploy LCD/LED display with the Micro PC, or (3) deploy Chromebook along with tablet. Option 2 leverages the benefits of low power consumption by both the LCD/LED display and Micro PC to provide users access to virtual desktops. Another benefit is the low cost of Micro PC. However, the cost of the LCD/LED display may be too expensive for developing countries. Option 3 may be a more cost effective option for developing countries based on the low price of each Chromebook and tablet along with the fact that these devices have both display and keyboard capabilities. In Option 3, users would access the virtual machines or desktops via WiFi using Chromebooks. Table 4 summarizes the cloud computing adoption strategy recommended for SMEs.

Besides the expected benefits from the micro cloud configuration, the proposed hybrid cloud configuration for SMEs also has the potential for dual-use benefits of

Table 4. Small and medium enterprise (SME) cloud computing adoption strategy

		Economic Level Dimension	
		Small and Medium Enterprise (SME)	**Example Configuration**
Adoption Characteristic Dimension	**Adoption Objective**	Offer cloud computing to support IT outsourcing so that SMEs can focus their core business operational efficiency and growth.	
	Cloud Type / Services	• Hybrid cloud • Public cloud services (SaaS, PaaS, IaaS)	• Open source hypervisor - Citrix XenServer • Hyper-converged servers - $$$
	Power Source	• Primary - National Grid - Mini/Micro Grid • Secondary - Backup generators - Alternative energy sources (e.g., solar power) - UPS w/ 30 minutes backup time	• Goal Zero Yeti 1250 Solar Generator - $1599.99 • APC Smart-UPS C 1500VA LCD UPS - 900 Watt - 36 min at 200 Watts - $310
	Client Device	• LCD/LED display with Micro PC • Portable devices - Chromebook - Tablet	• Micro PC - ZOTAC ZBOX PI320 pico - Atom Z3735F 1.33 GHz - 2 GB RAM - 32 GB SSD - Intel HD Graphics - Windows 8.1 - $169 • Chromebook ($150-$200) and tablet ($30-$150)
	Network Connectivity	LAN and WiFi with Internet / broadband connectivity scheduled for three times a day.	• Cisco Meraki Cloud Managed MS320-48 Switch - 48 ports - managed - $5,165
	Data Storage	• Primary - Local network storage • Secondary - Cloud storage	

supporting the United Nations SE4All initiative (e.g., providing lighting for medical operation, powering refrigerator to store medicines) (WHO 2015; Adair-Rohani, H. et al. 2013). More importantly, the SMEs' private clouds and available public clouds can serve as on-the-job training environment and employment opportunities. In fact, these should be close coordination between SMEs and local educational institutions (e.g., colleges, universities, technical trade schools) to ensure a growing pool of targeted, educated personnel resource pools. These industry-academic educational collaboration opportunities are prime candidates for international aid grants.

National / Regional

This chapter does not recommend a strategy for the national and regional level of developing countries because there is a wealth of literatures addressing the topic.

FUTURE PERSPECTIVES

From the technology perspective, cloud computing is rapidly evolving with advances in software products involving new level of virtualization to more efficient storage strategy. From the economic perspective, the projected adoption of consumer electronics in developing countries will experience the highest growth level in the coming years (Smeaton, 2016). In parallel, since 2013 there have been rapid gains in reported internet access rates in a large number of emerging and developing nations (Poushter, 2016). In this context, the cloud computing adoption framework for developing countries is a first step in the long journey in helping developing countries. To continue both the discussion and momentum of the proposed framework, there are some next step research areas described below that will refine and evolve the cloud computing adoption framework for developing countries:

- **The Economic of Hyper-Converged Infrastructure:** The area of hyper-converged infrastructure is a rapidly growing area with advances in both hardware and software integration to provide a turn-key solution for data centers. Hyper-converged infrastructure is one area where developing countries could achieve technology leapfrog in their cloud computing adoption strategy at the regional and national levels because of the high ratio of number of virtual machine capacity per hyper-converted server.
- **The Role of Software Container:** The wide spread adoption of software container technology, lead by Docker, is changing the way enterprise software is developed and deployed. Besides this context, the software container technology may offer a more efficient software model to support cloud computing for developing countries. Specifically, software containers are more efficient and responsive than virtual machines because they share the operating system or kernel whereas virtual machines have their own copy of the operating system. As software containers become available in more operating systems to include open source (i.e., free) operating systems, this trend provides developing countries more cost effective options in implementing micro cloud.

Figure 2. Cloud computing adoption strategy for developing countries

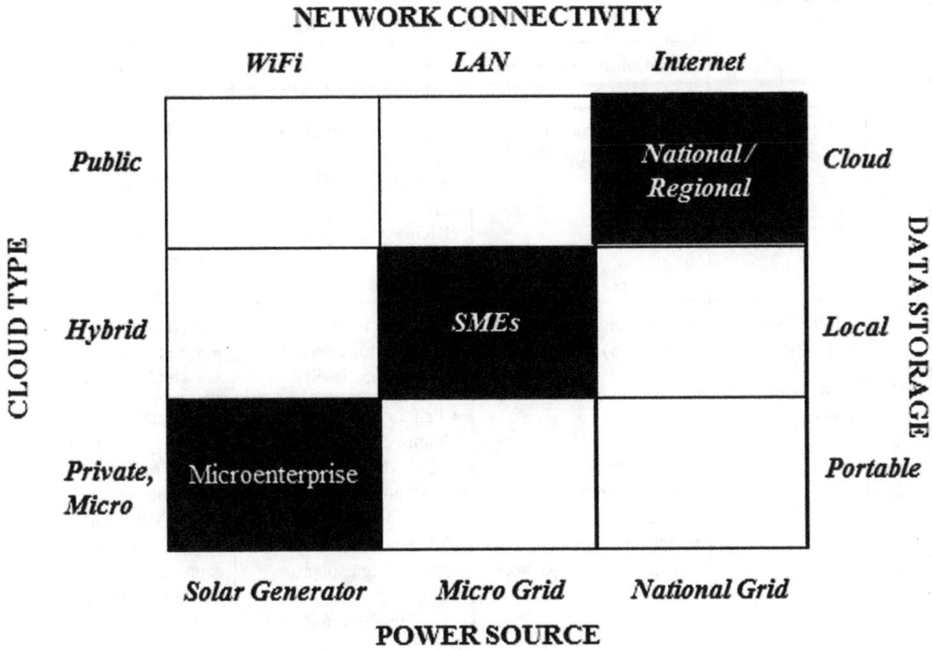

Figure 3. Deployment strategy for power source

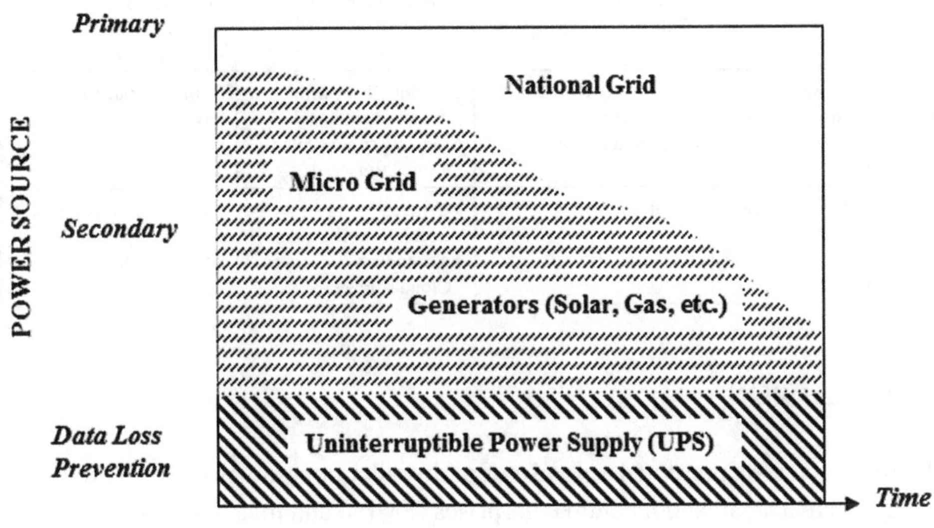

Table 5. Cloud computing adoption strategy for developing countries

		Economic Level Dimension		
		National / Regional	**Small and Medium Enterprise (SME)**	**Microenterprise**
Adoption Characteristic Dimension	**Adoption Objective**	Offer cloud computing as part of the national utility services.	Offer cloud computing to support IT outsourcing so that SMEs can focus their core business operational efficiency and growth.	Offer local population Internet café with virtual desktop to access computing resources to narrow the digital divide and enhance digital literately.
	Cloud Type / Services	• Public cloud (SaaS, PaaS, IaaS)	• Hybrid cloud • Public cloud services (SaaS, PaaS, IaaS)	• Private, micro cloud • Virtual machines • Virtual desktops
	Power Source	• Primary - National Grid • Secondary - Mini/Micro Grid - Solar power with battery backup - UPS w/ 1 hour backup time	• Primary - National Grid - Mini/Micro Grid • Secondary - Backup generators - Alternative energy sources (e.g., solar power) - UPS w/ 30 minutes backup time	• Primary - Solar power with battery backup • Secondary - Mini/Micro Grid - Alternative energy sources - UPS w/ 10 minutes backup time
	Client Device	• LED display with Micro PC • Portable devices - Chromebook - Tablet	• LCD/LED display with Micro PC • Portable devices - Chromebook - Tablet	• LCD/LED display with Micro PC • Portable devices - Chromebook - Tablet - Smartphone
	Network Connectivity	WAN and LAN with hourly Internet / broadband connectivity	LAN and WiFi with Internet / broadband connectivity scheduled for three times a day.	WiFi with Internet / broadband connectivity on an as needed basis.
	Data Storage	Cloud storage	• Primary - Local network storage • Secondary - Cloud storage	Local external personal storage device such as USB flash drive.

CONCLUSION

In this chapter, we presented a cloud computing adoption framework for developing countries that small and medium enterprises (SMEs) and microenterprises can use to enhance the economic growth of local industries. The framework incorporates proven strategies with the potential to quickly grow local industries, specifically adopting advanced technologies to narrow the productivity and research gaps. Figure 2 illustrates the recommended strategy for the different economic levels and major adoption characteristics of the cloud computing adoption framework for developing countries, i.e., an incremental adoption of cloud computing technology components – using available components instead of waiting for the availability of the entire solution.

Figure 3 illustrates the recommended strategy for power source in mitigating risks of unavailable or unreliable power infrastructure. The idea is to use UPS as a foundation and augment it with alternative energy sources such as solar power generator or micro grid to prevent data loss during a national power grid outage. The transition to the total reliance on the national grid can occur when the grid is reliable enough to support daily business operation without interruption.

Table 5 summarizes the recommended strategies for the cloud computing adoption framework for developing countries. The expected benefits of the proposed framework are: (1) providing a decision tool for cloud computing adoption addressing the different constraints at each level of the economy, (2) providing a context for cloud computing providers to customize their service offering to different segments of the economy and population, and (3) providing insights to formulate a holistic strategy for improving developing countries with coordination from various international initiatives that address only one aspect of assisting developing countries.

REFERENCES

Adair-Rohani, H. (2013). Limited electricity access in health facilities of sub-Saharan Africa: A systematic review of data on electricity access, sources, and reliability. *Global Health: Science and Practice*, *1*(2), 249–261. PMID:25276537

Adam, I., & Musah, A. (2014). *Small and Medium Enterprises (SMEs) in the Cloud in Developing Countries: A Synthesis of the Literature and Future Research Directions*. University for Development Studies. University of Ghana - Business School.

Bahaj, A. (2016). *Transforming rural communities through mini-grids*. Smart Villages. Retrieved November 28, 2016, from http://e4sv.org/transforming-rural-communities-through-mini-grids/

Comin, D., & Hobijn, B. (2010). *Technology Diffusion and Postwar Growth.* In D. Acemoglu & M. Woodford (Eds.), *NBER Macroeconomics Annual* (pp. 209–246). The University of Chicago Press. doi:10.3386/w16378

Comin, D., & Hobijn, B. (2012). How Early Adoption Has Increased Wealth – Until Now. *Harvard Business Review*, *90*(3), 34–35.

Comin, D., & Mestieri, M. (2010). *The Intensive Margin of Technology Adoption*. Harvard University and NBER.

Cottrell, R. (2013). How Bad Is Africa's Internet? *IEEE Spectrum*. Retrieved July 31, 2015, from http://spectrum.ieee.org/telecom/internet/how-bad-is-africas-internet

Demirgüç-Kunt, A. (2011). *Generating Jobs in Developing Countries: A Big Role for Small Firms*. The World Bank. Retrieved March 14, 2015, from http://blogs.worldbank.org/allaboutfinance/generating-jobs-in-developing-countries-a-big-role-for-small-firms

Foster, A., & Rosenzweig, M. (2010). *Microeconomics of technology adoption*. Center Discussion Paper No. 984. Economic Growth Center, Yale University.

Greening, J. (2015). *Helping developing countries' economies to grow*. Department for International Development. Retrieved March 14, 2015, from https://www.gov.uk/government/policies/helping-developing-countries-economies-to-grow

Grimshaw, D., & Lewis, S. (2010). *Solar power for the poor: Facts and figures*. Retrieved November 28, 2016, from http://www.scidev.net/global/energy/feature/solar-power-for-the-poor-facts-and-figures-1.html

Hashim, N., Murphy, J., Doina, O., & Connor, P. (2014). Bandwagon and leapfrog effects in Internet implementation. *International Journal of Hospitality Management*, *37*, 91–98. doi:10.1016/j.ijhm.2013.10.012

IEA. (2014). *World Energy Outlook 2014*. Paris, France: International Energy Agency.

Internet World Stat. (2014). *World Internet Users and 2015 Population Stats*. Retrieved July 31, 2015, from http://www.internetworldstats.com/stats.htm

ITU. (2012). *The Cloud: Data Protection and Privacy Whose Cloud is it Anyway?* GSR 2012 Discussion Paper. International Telecommunication Union (ITU).

Kapurubandara, M., & Lawson, R. (2006). *Barriers to Adopting ICT and e-commerce with SMEs in Developing Countries: An Exploratory study in Sri Lanka*. University of Western Sydney.

Kelly, A. (2013). Technology can empower children in developing countries - if it's done right. *The Guardian*. Retrieved July 31, 2015, from http://www.theguardian.com/sustainable-business/technology-empower-children-developing-countries

Lowe, S., Davis, D., & Green, J. (2016). *2016 State of Hyperconverged Infrastructure*. ActualTech Media. Retrieved November 29, 2016, from http://demand.simplivity.com/report-2016-state-of-hyperconvergence-thankyou

Manyika, J., Chui, M., Bughin, J., Dobbs, R., Bisson, P., & Marrs, A. (2013). *Disruptive technologies: Advances that will transform life, business, and the global economy*. McKinsey Global Institute.

Mell, P. & Grance, T. (2011). *The NIST Definition of Cloud Computing*. NIST Special Publication 800-145.

Miliaras, C. A. (2012). *Creating jobs that reduce poverty: A research agenda on developing-country gazelles. RTI Press publication No. OP-0011-1211*. Research Triangle Park, NC: RTI Press. doi:10.3768/rtipress.2012.op.0011.1211

Nobel, C. (2012). *How technology adoption affects global economies*. Harvard Business School Working Knowledge.

Oracle. (2013). *Oracle VM User's Guide for Release 3.1.1*. E27309-04. Retrieved July 31, 2015, from http://docs.oracle.com/cd/E27300_01/E27309/html/vmusg-virtualization.html

Pew Research Center. (2015). *Internet Seen as Positive Influence on Education but Negative Influence on Morality in Emerging and Developing Nations*. Retrieved July 31, 2015, from http://www.pewglobal.org/files/2015/03/Pew-Research-Center-Technology-Report-FINAL-March-19-20151.pdf

Poushter, J. (2016). *Smartphone Ownership and Internet Usage Continues to Climb in Emerging Economies But advanced economies still have higher rates of technology use*. Academic Press.

Practical Action. (2013). *Poor people's energy outlook 2013: Energy for community services*. Practical Action Publishing. Pew Research Center. Retrieved November 29, 2016, from http://www.pewglobal.org/2016/02/22/smartphone-ownership-and-internet-usage-continues-to-climb-in-emerging-economies/

SE4ALL. (2015). *Universal Energy Access*. United Nations Sustainable Energy for All initiative (SE4ALL). Retrieved July 31, 2015, from http://www.se4all.org/our-vision/our-objectives/universal-energy/

Smeaton, C. (2016). *6 New Ecommerce Trends Shaping the Future of Online Retail*. Demoup. Retrieved November 29, 2016, from https://www.demoup.com/blog/six-new-ecommerce-trends-shaping-the-future-online-retail/

Stanford. (2004). *Vinod Khosla: Microlending to End Poverty*. Insights by Stanford Business. Stanford Graduate School of Business. Retrieved March 14, 2015, from http://www.gsb.stanford.edu/insights/vinod-khosla-microlending-end-poverty

UNCTAD. (2013). *The Cloud Economy and Developing Countries*. Information Economy Report 2013. United Nations Conference on Trade and Development (UNCTAD). Retrieved July 31, 2015, from http://unctad.org/en/PublicationsLibrary/ier2013_en.pdf

WHO. (2015). *Access to Modern Energy Services for Health Facilities in Resource-Constrained Settings A Review of Status, Significance, Challenges and Measurement*. WHO.

Yeboah-Boateng, E., & Essandoh, K. (2014). Factors Influencing the Adoption of Cloud Computing by Small and Medium Enterprises in Developing Economies. *International Journal of Emerging Science and Engineering, 2*(4).

KEY TERMS AND DEFINITIONS

Cloud Computing: A model for enabling ubiquitous, convenient, on-demand network access to a shared pool of configurable computing resources (e.g., networks, servers, storage, applications, and services) that can be rapidly provisioned and released with minimal management effort or service provider interaction.

Extensive Margin: The time it takes a country to adopt a technology.

Hybrid Cloud: The virtual environment spread between both local virtual environment and remote virtual environment from Cloud provider.

Infrastructure as a Service (IaaS): A delivery method that provides users with servers and storage resources. An IaaS provider performs tasks such as system maintenance, backup and resiliency planning while the users is responsible for the application installed on the server.

Intensive Margin: The extent to which a technology is adopted by the population.

Platform as a Service (PaaS): A delivery method that provides users with server resources to develop or host applications. A PaaS provider hosts the required hardware and software on their infrastructure. As a result, users do not have to install in-house hardware and software to develop or run a new application.

Private Cloud: The virtual environment hosted within an enterprise and all services are available only via the enterprise network.

Public Cloud: The virtual environment hosted by a Cloud provider and all services are accessible from both the enterprise network and the Internet.

Software as a Service (SaaS): A delivery method that provides users access to application remotely. A SaaS provider performs the maintenance and support for the services while the users only have access to the account and the user's content.

Technological Leapfrogging: The process whereby an enterprise realizes several technological advancements bypassing standard technology and adopting modern technologies.

ENDNOTE

1 Dell. Youth Learning. http://www.dell.com/learn/us/en/uscorp1/power-possible-learning.

Chapter 9
ICT for Development in Nigeria:
Towards an Alignment With ICT4D 2.0 Goals

Olumide B. Longe
Caleb University, Nigeria

Adefolarin A. Bolaji
Purdue University, USA

Richard Boateng
University of Ghana, Ghana

ABSTRACT

Prior studies have x-rayed the potentials and impetus for development that can be resultant from a full scale adoption of ICT in Africa particularly in Nigeria. Current challenges relating to infrastructures, cybercrime, government policies and so on that mitigate the benefits accruable from a virile ICT growth have also been highlighted. However, research also needs to really address possible areas of developmental benefit of ICT to rural target groups such as farmers, extension officers, health and social workers This paper explores possible alignment of ICT4D2.0 goals to ICT development in Nigeria. It posited that the digital divide that ICTD1.0 sets to bridge are still very evident at grassroots level. Adding computing and Internet functionality to technologies that already penetrate the grassroots such as mobiles phones, radios and televisions will enhance ICT4D2.0 goals in Nigeria. This will in turn increase productivity among this target group with attendant contribution to national development.

DOI: 10.4018/978-1-5225-2565-3.ch009

INTRODUCTION

Information and Communication Technologies for Development (ICT4D) is a general term referring to the application of Information and Communication Technologies (ICTs) within the field of socio-economic development. "ICT4D" as a term is described as a strong and generally positive force. It helps in the magnetization and aggregation of knowledge and practice. It made provision for clear and unambiguous tag for searching events and material. The "4D" component indicates each activity's purpose. Without it, more loss than gain would be made (Richards, 2016).

ICT4D is geographically unspecific, and as such concerns itself directly with overcoming the barriers of the Digital Divide. It is becoming recognized as an interdisciplinary research field as can be noted by the number of conferences, workshops and publications in the field (UN-ESCWA, 2005; UICT4D, 2007 and SPIDER, 2008). Such research has been motivated by the need for empirical results that can be used to measure the impact of ICT4D initiatives. ICT is recognized by government, organizations, individuals and even criminals as a tool that can be used for varying degrees of achievements.

ICT4D can therefore be applied either in an indirect sense, where the ICTs assist aid organizations or non-governmental organizations or governments in order to improve socio-economic conditions or in the direct sense, where their use directly benefits the disadvantaged population in some manner. The former was more of the focus of ICT4D1.0 while ICT4D2.0 is expected to achieve the laudable goals of pragmatic and penetrative development at all levels. Unfortunately, government bureaucracy, lack of basic infrastructures, inadequate publicity, legislative and other political bottlenecks have hampered the rate of growth and penetration of ICT and its attendant development potentials in most developing countries.

On the one hand, developing countries present problems that are as diverse as the cultures they encompass, and the issues fully emerge only with direct interaction with the target communities. On the other hand, many of the issues are recurring, and they often are just minor variations of challenges encountered in the developed world (Kentayo and Benadi, 2008). Developers have identified the root of several ICT4D failures as stemming from their technocentric approach, dominated by an informatics view of the world. Such projects are frequently analogous with the old medical joke, "The operation was a success but unfortunately the patient died." They deliver a system that works technically but that fails to make a developmental contribution. To move away from the inadequacies of ICT4D 1.0, then, we must have new, broader worldviews guiding ICT4D 2.0 projects. These broader worldviews will likely come from two main disciplinary candidates: Information Systems (IS) and development studies (Longe, et al., 2007; Longe & Chiemeke, 2007a, 2007b and Richard, 2008). Increasing availability and access to ICT is very important if ICT4D2.0 is to make any meaningful contribution in Nigeria.

Development of ICT in Nigeria

The first Information and Communication Technology (ICT) initiative in Nigeria started in the 1950s with focus on print and electronic media. No major policy or other outcome was achieved because of strict government control. The full awareness of the importance of ICTs was absent. Only the private sector demonstrated ICT initiatives. The Federal Government of Nigeria has recognized the import of information technology as a major key to economic growth and sustainability. That culminated in a workshop on ICT policy in Abuja in March 2000 which brought together major ICT stakeholders such as the Nigerian Computer Society (NCS), IT Association of Nigeria (ITAN), the Computer Professionals of Nigeria (CPN) and other major stakeholders in the public and the private sector. These days, many conferences and workshops hold at different venues in Nigeria discussing the diffusion, penetration and growth of ICT in the nation.

With the collaboration of several committees, the ICT final draft of the National Information and Communication Technology proposal was approved by the Federal Government of Nigeria (National ICT Policy, 2012; Ahiakwo, 2002). According to its vision statement, the policy is aimed at making Nigeria an IT capable country in Africa and a key player in the information society by using ICT as the engine for sustainable development and global competitiveness. The mission statement, centres on using ICT for education, creation of wealth, poverty eradication, job creation, and global competitiveness. Unfortunately, there are no specifics about diffusion into the rural areas this seem to be a lofty goal and was over-sighted by the policy formulators.

Challenges Facing ICT Diffusion and Adoption in Nigeria

Information and Communication Technologies for Development (ICT4D) deals with the application of information and communication technologies to the development of programs in underdeveloped countries. Although some value-added ICT4D projects have been meticulously conceived and focused on some target group little has been realized in terms of achievements from these projects. The major challenges facing the adoption of ICT in Nigeria have been identified to include:

1. Absence of adequate communication network infrastructures.

2. Relatively high cost of ICT products that is not affordable to the large low-income Portion of Nigerians.
3. Relatively low level of support across the rank and file of people in government.
4. Problems associated with technical and management support for internet Connection.
5. Monopolistic tendencies among service providers and operators coupled with the problems of interoperability.
6. Lack of quality control over importation of ICT tools and equipment into the country.
7. Lack of foreign investors interests in doing business as a result of lack of basic infrastructures (Okyere et al., 2003).
8. The challenges of sustainable wired and wireless networks.
9. Cost of connection.
10. Security issues and fraud.
11. Political instability/policy inconsistencies and government bureaucracies.
12. Lack of effective coordination and other regulatory issues.
13. High running and subscription costs.
14. Lack of good publicity and incentives to attract potential users.
15. Identification of information sources that meet the needs of users.
16. Poor Quality of Service of the internet and telecommunication services.
17. Effective management of network traffic and infrastructure.
18. Lack of enabling environment for viable private sector activities.
19. Non awareness of software export potential.

BACKGROUND

Several authors have discussed the level of Internet and ICT penetration in Nigeria as very low, resulting from high cost of bandwidth, exorbitant prices for computer acquisition, and unreliable power supply (Ajayi, 2002; Anao, 2002; Nwanko, 2006; Olatokun, 2007; Longe, et al., 2007 and UN-ESCWA, 2005). A casual investigation of the Nigerian ICT terrain reveals that wireless network will grow faster than wired network. This fact is supported by the rapid increase in mobile telephone subscribers in recent years. Online banking and electronic commerce are also enjoying increases in embracement with the proliferation of ATM machines in notable and accessible points in different metropolis. This has however not come without the attendant problems of irregular and inconsistent services to customers, network problem and the issue of cybercrime (Longe & Chiemeke, 2007b, 2008; Omolola, 2005; Smith et al., 1999).

This is the age of ICT and the benefits of ICT for development cannot be underscored. ICT affects everybody, in everyday life and in every sector of the economy. According to World Bank Group Strategy for ICT (2012), ICTs have great promise to directly or indirectly lead to poverty reduction, productivity increment, economic growth boost, and accountability and governance improvement. The penetration of ICT into the society (Mutula, 2004) has transformed the society. According to Chacko (2005) ICT can contribute towards development. However, ICT alone is not sufficient to reach a country's development goals. The social aspects of information systems such as training are also important (Roode et al., 2004).

According to UNDP (2005), effective development of ICTs can affect human development in two ways. Firstly, ICT as a sector of economic activity can impact the overall economic growth of a nation. Secondly, ICT as an enabler for enhancing human productivity can influence human development through access to information knowledge and enlarging choices. However, Chacko (2005) argues that ICT helps in alleviation of poverty, enhancing education and improving healthcare.

Firm-level variables such as financial capacity and technological absorptive capacity influenced the intensity of the adoption of ICTs thereby suggesting that globalisation of the Nigerian economy also influenced the adoption of new technologies. Closing these gaps – locally, nationally and globally - requires creative pro-people policies that focus on national priorities – on areas that will have the greatest positive impact on development.

In Africa, some major problem facing the development of the full potential of ICT for education, research and development include brain drain which has resulted in the lack of the critical mass of ICT-engineers and scientists relevant for undertaking ICT-related project professionally and the non involvement and imbalance of gender and other target groups in policy formulation. Another major obstacle is the lack of an enabling environment and a sound ICT-roadmap and strategies by policy makers resulting in uncoordinated and unsustainable ICT-development activities (Joseph et al., 2003; Okyere et al., 2003; Olatokun, 2007).

Figure1 indicates the general overview of the technologies and processes of ICT4D. It gives a diagrammatic description of how innovation feasibility moves bottom-up from the basic infrastructure to the business model. User-Produced Services relates the Context to the Access model while User-Consumed services relates Access model to the Context (Richard, 2009).

Correlating ICT with Socio-Economic Development

One of the benefits of the ICT revolution are the opportunities now open to Nigerian businesses to access a wider global market through e-commerce. As e-commerce opens up the Nigerian market to foreign enterprises, the relatively affluent Nigerian

Figure 1. The technologies and processes of ICT4D (RIA, 2012)

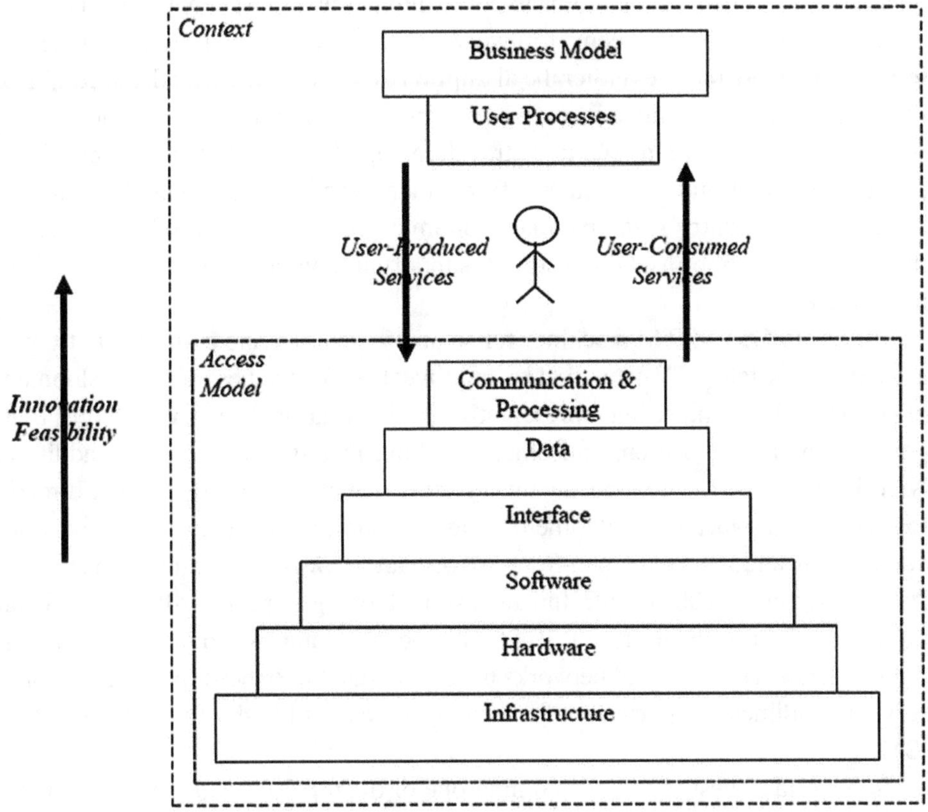

consumer with access to the internet is given much greater choice with regard to desired products and services. In contrast to the situation in the developed world, where transport and communications infrastructures for delivery of physical goods and information services are well established, the alternatives available in Nigeria are generally slow, expensive, and in some cases practically non-existent.

The communications and information delivery capability of ICT products and related services serves all sectors of society. Other sectors such as education, health, social policy, commerce, government, banks, trading, communications, science and technology have all benefited.

These resources are interlinked and synergistic; individuals can visit and exploit relevant information sources, which often point to additional sources of information and to knowledgeable individuals. The correlation between information, communication, and economic growth is well known, making the usefulness of networks nearly self-evident. Electronic networking is a powerful, rapid, and inexpensive way to communicate and to exchange information. With available networks, previously unanticipated collaboration seems to come into being almost spontaneously.

The underlying cause seems to involve a latent demand that remains latent as long as joint work requires either the disruption of waiting for the mail, the continual retyping of texts transmitted by mail or fax, or the need to secure large budgets and approvals for extensive international travel. In Nigeria networking is now crucial to scientific research and development efforts, many of which yield tangible economic benefits. The country's commercial/economic growth is enhanced by access to information and improved contact with support and purchasing personnel as well as customers.

Access to GSM, ATM machines for seamless banking and commercial/trading activities and satellite TV networks also improves the effectiveness of the development of communities, comprising representatives of international agencies, staff of non-governmental organization, and others working locally and abroad. In addition, Nigerian Universities are focusing on curricula that might contribute more directly to economic growth and development using collaborative efforts enabled by network connections and the Free Open-Source Software (FOSS) paradigm (Longe et al., 2004;Longe et al., 2008). Unfortunately, some target groups are still at the tail end of the benefits accruable from ICT4D – the poor and unreached in the rural areas. Although GSM systems and networks have penetrated into these spheres, they only serve the rudimentary purposes of calls and where possible short message services (SMS).

In Nigeria's quest towards becoming one of the top 20th economy of the world by the year 2020, Nigeria is expected to align with ICT4D goals which provides practical ways of diversifying the economy away from its over dependent on the oil sector. The people should be aware of the impact of ICT on their everyday life in terms of communication application: newsletters, websites, multimedia presentation; interactive communication applications: blog, wikis and social networking websites; data handling applications: surveys, school libraries, clubs and society records, shop records, address lists; measurement applications: scientific experiments, environmental monitoring and electronic timing; control applications: control of lights, automatic washing machine, automatic cookers, burglar alarm, central air condition controller, microwave oven, video recorder and player; modelling applications: 3D modelling, simulation, spread sheet for personal and shop finances (IGCSE ICT).

INFORMATION POVERTY AND ITS IMPLICATIONS

In Nigeria, information poverty is one of the more significant and insidious obstacles to effective exploitation of ICT4D. It has had the same effect on information processing and other types of technology. Lack of adequate information regarding development in other countries and other environments is often not noticed, and in the absence of new information, old techniques and procedures are continuously used without conscious knowledge of alternatives. This is more prevalent among grassroot service organizations involved in farming, poultry keeping, cattle rearing etc.

Other contextual constraints may be present (Mcnamara, 2003). In Nigeria, information poverty, financial poverty, and misperceptions about the costs and benefits of network connectivity have sometimes resulted in decisions to delay investment in ICT infrastructures which may be considered too expensive relative to other needs (Datta & Mbarika, 2006). For Educational and technical performance the ICT is a must because it will give all the information and connections we can possibly desire. Crowley and Ryan (2013) identified sustainability and scale, lack of knowledge, changing roles and norms, pace of change, and funding as key challenges and possible threats of implementing ICT4D.

ALIGNING ICT DEVELOPMENT IN NIGERIA WITH ICT4D2.0 GOALS

During ICT4D 1.0's debut, developers rapidly recognized that plugging a peasant farmer or slum-dweller into Google offered limited value. Much of the information they required would not emerge because it was not in digital format. A series of projects, such as Open Knowledge Network, sought to create relevant local data content focused on livelihood-appropriate issues such as health, education, agriculture, and rights. Once media technologies like radio and television were incorporated into ICT4D, developers recognized that its non-interactive and broad-scale nature presented a lack of specific data relevance. For these technologies, the phase change to ICT4D2.0 has been associated with community radio—much localized broadcasting that allows community input. Community television is not yet a realistic prospect, but its equivalent, participatory video, provides for the creation of video content by the local community, presented at individual screenings for community groups (Richard, 2008).

Although access figures are hard to come by, about 80% of the population in developing countries has access to a radio and about 50% to a television. Early in ICT4D's history, these statistics prompted the swift re-interpretation of ICT to incorporate radio and television, and foreshadowed the role convergence would play in ICT4D2.0. Looking at the technologies that already penetrate—mobiles, radios, televisions— developers must now seek ways to add computing and internet functionality to these tools. This must be the quest in the developing world such as

Nigeria where convergence will definitely play a role in aligning with the ICT4D2.0 (Kentayo & Benadi, 2008; Longe & Chiemeke, 2007a).

Music plays a big role in advertising industry in developing countries, it has been used greatly in creating awareness in politics, census, new products and UNDP development programmes like immunization in both Urban and Rural areas. Nigeria can leverage on ICT music, ICT dance steps and ICT slogans for ICT penetration in Nigeria. Music in advertising affects the way viewer perceives the brand. According to Huron (1989), advertising is the means by which one party attempts to convince or entice another into purchasing a particular product or service. According to Jaguda (2013), music stays in people's mind in developing countries. ICT can leverage on music to create awareness about impact of ICT on everyday life. Nigeria has used music to disseminate information during census and Ebola disease crisis for urban and rural dwellers. According to the RIA Nigeria 2012 ICT Survey, one of the principal activities conducted with the mobile handset by individuals is listening to music (55%).

According to the Internet World Statistics (2016), the number of internet users in Nigeria as at year 2000 was 200,000 however, by June 2016, the number of internet users has grown to 92.7 million users out of the 181 million population. By December 2012, number of Facebook users in Nigeria is 6,630,200 and by November 2015, it was estimated to be 15,000,000; this points towards the alarming rate at which ICT usage is increasing by the day.

New ICT-related tools can make institutions and markets more productive, enhance skills and learning, improve governance at all levels, and make it easier for the poor to access services and make their voices heard. Beyond the internet, which seems to have greater acceptance and penetration in the big cities, Nigeria must exploit cheaper technologies already in use to deliver sufficient ICT functionality to make a difference. Rather than wait for handset and bandwidth upgrades to allow mobile internet access, we must determine what can be achieved for development through calls and SMS and, possibly, older technologies. Interface innovation that can drive access to ICT-based information, services, education, instructions and jobs for all local languages must be devised. Investors must be encouraged and motivated to tap into these opportunities that will be of mutual benefit to them and the country at large.

At the rural level, techniques must be inculcated into already existing tools that can boost crop yield, advice on fertilizer application, improve well-drilling, teach mothers how to better care for infants and give health specialists current information to effectively coordinate and halt the spread of diseases, provide information on weather forecast and news of current development applicable to dairy and cattle rearing/production. Patients can be reminded of the need to take their medications from local health centres and maternity with SMS or text message and nursing mothers ca be given awareness for vaccinations through text messages.

Another area of aligning ICT Dcvclopment in Nigeria with ICT4D2.0 goals is the area of security. People have been protecting their properties for generations with locks, seals, fences, signatures, meters and account books, this is changing rapidly to electronic form, several transactions and shopping are being moved to the internet and this development has not left out Nigeria. Unfortunately, security breaches on the internet are also gaining wide popularity with the emergence of the electronic forms of security (Ross, 2008).

Banks in Nigeria have suffered a lot of ATM fraud because of security breaches. Individuals have experienced terrible shock as a result of financial transaction session hijacking and private information sniffing over the internet. In aligning with the goals of ICT4D, formidable security models should be built into the security structure of the country, the use of technology should be fully embraced at all levels of the government be it local, state or federal. Security experts and analysts are to be employed and further trained in every public and private organization. This will reduce the proliferation of terrorism and fraudulent activities in the entire system of the country. ICT tools should also be made use of as a platform for educating the citizen on personal and corporate security using mobile applications, social networks, official and non-official gateways.

CONCLUDING REMARKS

The solution strategy towards bridging the digital divide demands an aggressive human capacity building in ICT through training workshops, seminars and courses in collaboration with local and international institutions. Convergence of already existing neo-ICT tools and medium such as radio and television in addition to penetrable modern ICT tools such as mobiles phones will also assist in realizing ICT4D2.0 goals. Borrowing from the experiences of other African countries and the advanced nations, bodies set up for the implementation of ICT4D2.0 goals in Nigeria must be strengthened and research and development must be demand-driven focusing on the provision of products to meet local needs.

ICT development is multidimensional. The Public sector must partner with private sector in a creative and mutual manner. Ideas are not enough, policies are not enough, and this requires close co-ordination, coherence for Nigeria to use ICT to drive socio and economic development.

There is also the need to deviate from technocentricity in adult ICT literacy classes. We must begin to see ICT and computer literacy in terms of placing in the hand of learners an all-encompassing ICT tool and knowledge that will equip them with the ability not only to operate computers, but also to use any computing and allied facilities. Those who teach adults computer technology should approach it with a great deal of pedagogical responsibility. It is important that educators involved in using computers for instruction or as a subject of instruction be cognizant of the computer as more than "just a tool." It is a cultural artefact with meaning, a meaning deeper and richer than the computer's mere physical presence.

To align with the ICT4D2.0 goals stakeholders must do the following:

1. Educate the government on the need to formulate policies that will aid the use of ICT tools for enlightenment programs that among rural dwellers on issues that directly affect their welfare such as farming, health, productivity and social awareness.
2. Network operators should provide incentives for users and other agencies that enable better and more pragmatic usage of their infrastructure to aid in ICT diffusion in the rural areas.
3. User-friendly features and instruction to aid the low learners must be inculcated into ICT imported into our nation.
4. Incentives such as reduction in calls tariff and internet access through phones should be directed at the grassroots.
5. Teachers of technology should go beyond the usual "text" of instruction manuals and online help menus, and explore and explicate a fuller meaning and understanding of computer literacy involving practical demonstration of cutting edge ICT technology and theory applications (Longe, 2007).
6. Local software developers should be encouraged to produce local content for local organisation. Software developers should harness their potentials for local software market.
7. Awareness of export potential of local software. Due to devaluation of Nigeria currency, it's cheaper for foreigners to buy software from Nigeria. Nigeria can export software developed by local people and earn in foreign currency.
8. ICT for everyday life awareness through ICT music, ICT dance steps and ICT slogans.

REFERENCES

Ahiakwo, C. O. (2002). *The Role of Internet Connectivity in Nigeria*. Available Online at: http://www.isocnig.org.ng/conference/paper

Ajayi, G.O. (2002). *Challenges facing the Nigerian Telecoms Industry*. Address Delivered at: Nigeria Telecoms Summit Abuja.

Anao, A. R. (2002). *Positioning Nigeria for effective response to the Challenges of Emerging Technologies and Globalisation*. Available Online at: http://www.globalpolicy.org /globaliz/econ/2002/124ictafrica.htm

Chacko, J. (2005). Paradise lost: Reinstating human development agenda in ICT, policies and strategies. *Information Technology for Development, 11*(1), 97–99. doi:10.1002/itdj.20005

Crowley, J., & Ryan, M. (2013). *Building a Better International NGO: Greater than the Sum of the Parts?* Lynne Rienner Publishers, Inc.

Datta, P., & Mbarika, V. (2006). A comparative global investigation of casual relationship between information infrastructure and service sector productivity. *The Information Society*, *22*(3), 149–163. doi:10.1080/01972240600677847

Huron, D. (1989). Music in advertising: An Analytic paradigm. *Musical Quarterly*, *73*(4), 557-574. Available Online at: www.ohio-state.edu/Huron/Publications/huron.advertising.text.html

IGCSE ICT. (n.d.). *ICT Use in Everyday life*. Available Online at: www.igcseict.info/Theory/7_1

Internet World Statistics. (2016). Available at: http://www.internetworldstats.com/stats1.htm

ITU. (2005). *Access indicators for the Information society*. Tunis: World Summit on the Information Society.

Jaguda. (2013). *8 ways you can move from An Unknown Artiste to a Renowned Artiste in Nigeria*. Available Online at: www.jaguda.com/2013/05/22/8-ways-you

Joseph, I., Philip, F., & Shiloh, O. (2003). *Use of ICT for education, research and development in Ghana: Challenges and potentials. Round table on developing countries access to scientific knowledge*. Trieste, Italy: The Abdus Salam ICTP.

Kentayo, M. & Benadi, D. (2008). *Information and Telecommunication Technology for Development*. IEEE Positon Paper.

Longe, O. B., & Chiemeke, S. C. (2007a). Information and Communication Technology Penetration in Nigeria: Prospect, Challenges and Metrics. *Asian Journal of Information Technology*, *6*(3), 280–287.

Longe, O. B., & Chiemeke, S. C. (2007b). Sifting Paradigm in the Antispam Effort-SPAMAng: A Domain specific Outbound filtering for dealing with Nigeria Fraudulent mails. Journal of Computer Science and Information Technology.

Longe, O. B., & Chiemeke, S. C. (2008). Mediated Cyber- Crime: An Investigation into the Role of Internet Access points in the facilitation of Cyber Crime in South Western Nigeria. European. *Journal of social sciences*, *6*(3), 466–472.

Longe, O. B., Chiemeke, S. C., Onifade, O. F. W., Aliga, P. A., & Longe, F. A. (2008). Acceptance of the Open-Source Software Development Paradigm Among Undergraduates. *Proceeding of Open Source Software IDLEO Conference.*

Longe, O. B., Onifade, O. F., Chiemeke, S. C., & Longe, F. A. (2007). User Acceptance of Web-marketing in Nigeria: Significance of Factors. *Proceeding of the International Conference on Applied Business and Economics.* Available Online at: www.icabeconference.org

Longe, O. B., & Otti, V. U. (2007). Techno Phobia and its Impact on Adults Learning to use Computers in South western Nigeria. Journal of Information Technology *Impact,* 7(1), 81–90. Available Online at www.jiti.net

Longe, O. B., Uwadia, O., & Omoruyi, I. (2004). The Economic and Legal Implications of the open- Source Software Alternative.*Proceedings of 19th National Conference of the Nigerian Computer Society.*

McNamara, K. S. (2003). *Information and communication technology, poverty and development: Learning from experience.* Washington, DC: World Bank.

Mutula, S. M. (2004). Making Bostwana An Information Society. Current development. *The Electronic Library*, *22*(2), 144–153. doi:10.1108/02640470410533407

National, I. C. T. (2012). *Policy*. Ministry of Communication Technology.

Nwankwo, J. (2006, July 19). Computlife. *The Guardian*, 33.

Okyere, P. F., Intsiful, J. D. K., Osae, S., Allotey, F. A., & Andam, K. (2003). *Project proposal establishment of an african institute of science and technology in Ghana.* Trieste, Italy: ICTP.

Olatokun, W. M. (2007). Availability accessibility and use of ICT among Nigerian women Academics in Nigeria. *Malaysia Journal of Library and Information Science*, *12*(2). Available Online at www.ejum.org

Omolola, A. (2005, January 4). Cyber-fraud leads to blockage of Nigeria's Internet protocol Addresses. The Punch Newspaper.

Research ICT Africa (RIA). (2012). *Household and small business access & usage survey*. RIA Survey Methodology Brief. Available at: www.researchictafrica.net/publications/ Research_ICT_Africa_Policy_Briefs/2012_Stork_-_ICT_Survey_Methodology.pdf

Rhoode, D., Speight, H., Pollock, M., & Webber, R. (2004). It is not the digital Divide- it's the socio-techno divide. *Proceeding of the 12th European Conference on Information Systems*.

Richard, H. (2008). ICT4D 2.0. The Next Phase of Applying ICT for International Informatics Development. University of Manchester.

Richard, H. (2009). ICT4D 2.0 manifesto Where Next for ICT and International Development? Development Informatics Groups, Institute for Development Policy and Management, Manchester, UK.

Richard, H. (2016). *From ICT4D to Digital Development?* Available Online at: https://ict4dblog.wordpress.com/page/2/

Ross, A. (2008). *Security Engineering: A Guide to Building Dependable Distributed Systems*. Wiley Publishing Inc.

Smith, R., Holmes, M. N., & Kaufmann, P. (1999). *Nigeria Advance fee fraud, Trends And Issues in Crime and Criminal Justice, 121*. Australian Institute of Criminology Canberra. Available Online at: http:/www.aic.gov.au

Spider. (2008). *Swedish Programme for ICT In Developing Regions*. KTH.

UICT4D. (2007). *University ICT4D*. University of Washington.

UN-ESCWA. (2005). *United Nations on Partnership on measuring ICT for Development*. UN-ESCWA Report on core ICT Indicators.

UNDP. (2005). *Human Development Report*. United Nation Development Programme.

World Bank Group Strategy for ICT. (2012). *ICT for Greater Development Impact*. Available Online at: https://siteresources.worldbank.org/EXTINFORMATIONANDCOMMUNIC ATION ANDTECHNOLOGIES/Resources/WBG_ICT_Strategy-2012.pdfwww.ncs.org.ng

KEY TERMS AND DEFINITIONS

ICT: Information and Communication Technology.
ICTs: Information and Communication Technologies.
ICT4D: Information and Communication Technologies for Development.

Chapter 10
Level of Computer Appreciation Skills Acquired for Sustainable Development by Secondary School Students in Nsukka LGA of Enugu State, Nigeria

Chijioke Jonathan Olelewe
University of Nigeria – Nsukka, Nigeria

Igboamalu Peter Nwafor
University of Nigeria – Nsukka, Nigeria

ABSTRACT

This study aimed at determining the level of computer skills acquired by secondary school students in Nsukka Local Government of Enugu State, Nigeria. Specifically the study sought to ascertain the level of word processing skills, graphic application skills and Internet skills acquired by secondary school students in Nsukka Local Government of Enugu State. Three research questions and three null hypotheses were formulated in line with the objectives to guide the study. The study employed a descriptive survey research design using a sample of 272 respondents drawn from three secondary schools selected for the study. The instrument used for data collection was a 41-item structured questionnaire with a reliability index of.72,.83, and.80 using split half method. Data collected was analyzed using mean and standard deviation for research questions while t-test statistic was used to test the null hypotheses at 0.05 level of significance. The study revealed that word processing

DOI: 10.4018/978-1-5225-2565-3.ch010

skills and Internet skills are highly acquired by students' whereas graphic skills were moderately acquired by the students. It was therefore recommended that all education stakeholders should evolve supporting strategies like donating one dollar laptops to schools, retraining programmes and counterpart funding that will enhance computer literacy in Nigerian secondary schools.

INTRODUCTION

The goal of any meaningful education is to promote sustainable manpower development. Sustainable development means the development that continually meets today's need of any nation in a way that does not jeopardize the future generation (Pietella in Aigbovbiosa, 2006). This implies that sustainable development of a nation means the ability of that nation to maintain its economic, political, social and technological growth over time in a way that ensures the well being of that nation at that present time and in the future. According to UN (2008), sustainable development is defined as development that meets the needs of the present without compromising the ability of future generations to meeting their own needs. By this definition it is implied that sustainable development is a process that is continually evolving. The two major key concepts underpinned this definition: the concept of need and the idea of limitations imposed by the state of technology and social, political and economic as well as some human behavior and practices to meet present and future needs.

According to Okporie (2014), a development path that is sustainable in a physical sense could theoretically be pursued even in a rigid social and political setting. The author, however, pointed out that physical sustainability cannot be assured unless developmental policies pay attention to effective policy implementation and regulation approaches to socio-economic issues and practices which hinder development and its sustainability. The United Nations report (UNDESA, 2008) showed that, sustainable development has remained elusive for many African nations. For instance, poverty is still a major challenge, as 41% of sub-Sahara African populations (about 300 million people) are still living on one dollar per day or less (www.un.org/esa/sustdev/sdissues.htm).

Thus, one major means of overcoming grinding poverty and hunger is through functional education which guarantees meaningful source of livelihood through paid or self employment. Functional education is a type of education that is characterized by an intensive approach in contrast to the extensive approach in traditional literacy programmes. It is an education process that focuses on helping individual students develop saleable skills (21st century skills as ICT literacy skills, technological literacy skills, media literacy skills, among others), values, attitudes and knowledge required in the knowledge driven economy to become productive members of the society (NCREL & Metiri Group, 2003; Allen & Van der Velden, 2012; Voogt & Pareja, 2010; Pacific Policy Research Center, 2010). Today, the educational systems around the world are under increasing pressure to use Information and Communication Technology (ICT) to teach students the knowledge and skills needed in the 21st Century. ICTs are therefore electronic technologies used for accessing, processing, gathering, manipulating and presenting or communicating information. Information and Communication Technologies (ICTs) continues to impact positively on every aspect of human existence, thereby creating a powerful force for changes on how people conduct their daily businesses and in fact, determine the status of the Nation (Olelewe & Nzeadibe, 2015).

The advancement in technology has created so many ICT media that are necessary and useful in the teaching and learning processes. These ICT media consists of computer, television, Internet, GSM (Global System for Mobile Communications) and other media for collection, storage, processing, transmission and presentation of information (voice, data, texts and images) (World Bank, 2002). Today, the world is in the age of information technology transfer which is facilitated by the use of computer and ICT resources. The imperative of computer and modern ICT resources for classroom teaching and learning is further bolstered by the fact that these students are now "digital natives" who are born into the ICT culture and therefore can best be groomed using ICT tools. This view was supported by Wokocha (2013) who asserted that a good number of students are quite familiar with the world wide web, search engines, use of social media e.g. Facebook, Twitter, Yahoo Messenger, LinkedIn, YouTube, Watsapp, 2go, Skype etc such that in a single handset, they have over 100 applications running live thus informing and entertaining them.

Furthermore, computers like iPad, laptops, palmtops, desktops etc are increasingly widespread, influencing many aspects of our social and work lives, as well as many of our leisure activities. As more tasks involve human computer interaction, computer skills and knowledge have become more positively correlated with both occupational and personal success (Alabi, 2001). Bakpo (2005) defined computer as automatic electronic machines that can accept data, store the accepted data in its memory; manipulate the data according to instructions stored in the computer's memory and produce intelligible results from the manipulated data. Computer according to Gbadeyan (2005), may be defined as an electronic device which accepts and processes data by following a set of instructions (programmes) to produce accurate and efficient results at high speed. Gaining an appreciation of the computer by teachers and students may provide useful insights into technology integration, acceptance and usage in teaching and learning processes.

Computer appreciation according to Shapiro & Hughes (1996), is viewed as a new liberal art that extends from the knowledge of how to use computers to access information to critical reflection on the nature of information itself. Candy (2000) defined computer appreciation as the ability to manage and use computers for problem solving, decision-making and research. According to Ajayi (2000), computer appreciation is the knowledge and ability to use computers and related information and communication technologies efficiently, with a range of skills covering levels from elementary use (such as booting and shutting down of computers) to advanced problem-solving using computer application packages like Microsoft Office suits, Video-conferencing systems, database systems, graphical packages like CorelDraw, Adobe PageMaker, Webpage development programs like Dreamweaver, WebCT, Macromedia Flash, among others. Achuonye (2003) viewed computer appreciation as the level of ICT skills one has mastered while using computer programs and other applications packages.

Today, computer appreciation is the first step for any beginner in the use of Information and Communication Technology tools. Computer appreciation starts from knowing the basics of the computers like the components (Central Processing Unit, Visual Display Unit, Keyboard, Mouse and Speakers), booting of computer, ability to use application packages like Microsoft Word, Microsoft Excel, Outlook, Word Star and Word Perfect), Graphic packages (Corel Draw, Print master, Harvard, AutoCAD etc), Database packages (Dbase III, Dbase IV, Paradox, Clipper, FoxBase), Spreadsheet packages (Lotus 1-2-3, SPSS, Excel etc), Desktop publishing packages (Gold, Ventura publisher, Dbase and Adobe PageMaker etc), Anti-Virus packages (Norton Anti-Virus, Norton disc doctor, McAfee, Dr. Solomon's Took Kit, Penicillin, Avast, Avira etc), Games packages (Chess, Scrabble etc), Communication packages (Data talk, Cross talk and Data Soft), Electronic reference (e.g. Encarta, World Book), Presentation software (e.g. PowerPoint), hypermedia, email, Internet, (www) and every other basic knowledge of the computer including Assistive technologies like iMovie, iPad, iPod, tablet PCs and other mobile devices. After mastering computing skills at this level, one is expected to work with computers independently with little or no supervision and produce meaningful outcome that is rewarding. However, some people may decide to relent at this stage which they consider enough for them while others out of curiosity may move on to acquire more skills in more detailed form. Acquiring computer appreciation skills has given many school leavers across the globe a means of livelihood whether in industry, commerce, agriculture or micro-enterprise (King, 2005; World Bank, 2004; King & Palmer, 2007). Researcher have shown that by the year 2025, about 85% of the world population will be able to acquire household computers just like television in the families all over the world (World Bank, 2002).

Looking around the world today, it can be argued that computer has really made huge impact in the way people live their lives (Sheldon, 1999). Life is more simple and flexible with the computer; quick decisions are made where necessary and above all, the whole world has been made a global village through it (McLuhan & Bruce, 1989). Recently, the idea of computer appreciation has really permeated all areas of human endeavours and drawn people's attention, thus, having basic computer knowledge and skills has become prerequisite for any paid employment, admission into universities, and even basis for travelling to most advanced countries.

Skill as a concept is the ability to use human knowledge effectively in carrying out a given task. Skill is the ability to carry out a task with pre-determined results often within a given amount of time, energy, or both (Howland, 2013). Skill is perceived as the manual dexterity acquired through repetitive performance of an operation. Obi (2005) defined skill as the ability to use one's knowledge effectively and readily in performing an act. A person is said to have acquired skill when he can finish a given piece of work at a given time with minimum errors. To achieve optimum productivity in whatsoever occupation an individual is engaged in requires specialized skills. Skills are therefore regarded as an enabler for quality service delivery. Possession of high-level skill is necessary to enable school leavers utilize ICT facilities at their disposal (AMCIS, 2011).

There are many skills to be learned in every field. Offorma (2002) classified skills into intellectual, social (inter-personal and inter-group), and psychomotor skills. Intellectual skill has to do with the brain as it concerns reasoning. It progresses through knowledge level where students are expected to recount facts to evaluation level where the students are expected to pass judgments, value statements and take decisions based on internal evidence and external criteria. Social skill concerns with the intended outcome that has to do with attitude, character, emotion and temperament of the students. It progresses through receiving level where students tries to modify or adjust their behaviours to characterization level where affection, love or attitude are manifested maximally. Psychomotor deals with the manipulative or practical skills acquired by the students. It progresses through perception level where students are just being introduced or made to be aware of the phenomenon-skill through motion pictures or book study to origination level when students are able to use the skills they have acquired to bring about innovations and breakthroughs that are unique to them (Osuala, 2004). The development of each of these skills varies with the nature, complexity and the type of activity. Therefore, people who opt for skills training should among others possess aptitudes, patience, personality characteristics, and other human or physical qualities that will enable them succeed in performing any operation (Osinem, & Nwoji, 2005). Thus, educators and students must possess computer skills if they are to grow professionally, remain functional and relevant citizens.

Computer skills refer to the ability to use the software and hardwarc of a computer. Shapiro and Hughes (1996) defined computer skills as the practical skills involved in effective use of information technology and other electronic information resources. Computer appreciation skills are the ability to recognize when information is needed and have the ability to use the computer to locate, evaluate and use the needed information effectively (Grass Root Design, 2004). Research by Zhao and Alexander (2002) on technology and computer skills needs of students identified 28 computing skills including hardware skills such as using keyboard, microcomputer, and mouse, operating system skills such as Windows, productivity skills such as using word processor, spreadsheet, presentation, and database; with Microsoft office as the "strongly recommended" productivity application suite. Also, identified includes telecommunications skills namely using email, Internet/Web, intranet, wireless/mobile Web applications, Web-conferencing, searching/downloading/managing/posting information and information systems skills such as budgeting, general ledger, cash management, data mining, sales analysis, sales forecasting, decision support system, human resource management, systems analysis and design, and systems implementation. These skills should be identified by teachers and if properly taught as part of computer appreciation course it will help to equip secondary school graduates' with requisite skills needed to either secure a job or create one for himself and others. According to Eisenberg & Johnson (1996), delineating basic computer literacy skills is one way of ensuring that all aspect of computer literacy is achieved by secondary students.

Computer appreciation skills according to Weaver (2000), relate to the ability to recognize and use application packages such as database, spreadsheet, word processing, Excel, PowerPoint, data processing and Internet to effectively and efficiently accomplish a specific purpose. It is the basis for life-long learning. It is assumed that every student who passed through secondary education should have access to computers and Internet, having acquired the relevant skills needed to effectively use it.

Today, computer appreciation skills are taught at all levels of education in Nigeria: primary, secondary and tertiary education. In Nigeria, the federal and state governments own and manage public secondary schools through the federal and state ministries of education. This implies that the government provides funds, personnel and management for the operations of the public secondary schools in the country while the individuals or corporate organisations own and manage private secondary schools. Secondary education according to Federal Ministry of Education (2004), is the education children receive after primary education and before the tertiary stage. It is the formal education for children between ages 13 to 18years (Torane, 2010). According to Omeregie (2005), it is the point from where all the boulevards of higher education take off and all primary school leavers must pass through to become useful to themselves and the society. Secondary education occupies a strategic position in the education process because of the important role it plays in preparing individuals for useful living in the society and for higher education (Ejionueme, 2007). Secondary education enable children discover their strengths and weaknesses through different activities and assimilation (Obi, 2005; Pollak, 2009; Smith, Cowie & Blades, 2003; Taggart, 2007). The training in secondary education will also help shape their dreams and build their character to become responsible citizens of the society. Recently due to the failing economy more and more children are deprived of higher education (Babalola, 2007; United States Embassy in Nigeria. 2012; Oyaziwo, Philipa, & Imonikhe, 2012). It is on this note that secondary education should be made to provide enough opportunities for these children through function education to enable them support themselves, their families and at the same time contribute to nation building (Torane, 2010). This can be made possible if students are provided with facilities and relevant skills imparted to them on how to perform different functions using computers and other relevant ICT facilities.

Computer appreciation was integrated into the 9-year Basic Education Curriculum (that is, 6-year primary education and 3-year junior secondary education) by the Nigerian Educational Research and Development Council (NERDC) in the year 2007; hoping that the teachers and students for whom the curriculum are produced would demonstrate commitment and assiduity in using it. It was aimed at equipping these children with knowledge and skills that will enable them to secure job opportunities even when they are deprived of higher education.

Considering the advantages of acquiring computer skills by students in their educational pursuit, a question is posed, "are secondary school students in Nsukka Local Government Area of Enugu State among the beneficiaries of this veritable tool and new instructional possibilities"? If the answer is yes, what level of computer appreciation skills have they acquired in order to advance in their studies, secure job placement and become responsible citizens of the society in this information driven age? These questions prompt the researchers to investigate the level of computer appreciation skills acquired by secondary school leavers in Nsukka Educational zone with a view to provide appropriate recommendations towards teaching and learning of computer studies in Nigerian secondary schools.

In spite of the various usage of computers in teaching, learning, research and communication, the level at which computer skills are utilized in secondary schools is yet to be determined. Premised on the foregoing, this study intends to determine the level of computer appreciation skills acquired by secondary school students in Nsukka Local Government of Enugu State.

STATEMENT OF THE PROBLEM

One of the reasons for ICTs utilization in the classroom is to adequately prepare our students for a workplace where ICTs, particularly computers, the Internet and related technologies are becoming more and more ubiquitous. Computer has been identified as the major ICT tool for without it no other computer-based tools or facilities can function on its own basically for effective application in teaching and learning process.

Computer appreciation has been integrated into the 6-3-3-4 system of education Nigeria (that is, 6-years primary education, 3-years junior secondary education, 3-year senior secondary education and 4-years tertiary education) with the hope that teachers and students for whom the curriculum are meant for will demonstrate commitment and assiduity in using the curriculum (Federal Ministry of Education, 2007). The introduction of computer education at all levels of schooling was therefore aimed at producing quality graduates through the use of relevant curriculum which incorporates computing skills in areas of word processing, data processing, spreadsheet processing, database management skills, graphics and Internet skills, among others.

Unfortunately, due to current economic recession in Nigeria, young people are faced with numerous challenges ranging from unemployment, extreme poverty and hunger as more of them are unable to acquire higher education (Babalola, 2007; United States Embassy in Nigeria. 2012; Oyaziwo, Philipa, & Imonikhe, 2012). Some find it difficult to complete secondary education thus leading to high incidence of social vices amongst jobless youth. Statistics has shown that these group of young and energetic youths constitute 80million out of the 140 million Nigerians (NPC, 2006), representing 60% of the total population of Nigerians with 64million (80%) of them unemployed while 1.6million who are underemployed poses greater danger in terms of sustainability of the Nigerian economy. Similarly, youth unemployment is seen to be highest among secondary school leavers (NBC, 2012) with less than half being able to proceed to higher education. Similarly, studies have attributed high rate of graduate unemployment to lack of 21st century skills such as computer literacy skills, ICT skills, technological literacy skills, media literacy skills, among others (NCREL & Metiri Group, 2003; Allen & Van der Velden, 2012; Voogt & Pareja, 2010; Pacific Policy Research Center, 2010). Through the acquisition of relevant knowledge and computing skills, these young Nigerians will be able to secure job opportunities for themselves or become self-employed and thus reduce the high incidence of youth unemployment associated with armed robbery, rape, kidnapping, illicit drug trade and addiction, prostitution, terrorism as well as youth restiveness.

Hence, the introduction of computer education and ICT at all levels of education and training in Nigerian for over a decade now call for an appraisal of how well students have benefited from the subject for socio-economic empowerment. It is therefore pertinent to appraise the level at which the objectives of the programme have been attained. It was on this basis that the researcher situated this study to ascertain the level of computing skills acquired by secondary school students in Nsukka Local Government of Enugu State. In view of this, the researchers formulated three research questions aimed at addressing the purpose of the study:

1. What is the level of word processing skills acquired by secondary school students in Nsukka Local Government Area of Enugu State?
2. What is the level of graphic application skills acquired by secondary schools students?
3. What is the level of Internet skills acquired by secondary school students in Nsukka Local Government Area of Enugu State?

HYPOTHESES

Three null hypotheses were formulated to guide the study and were tested at 0.05 level of significance.

Ho_1: There is no significant difference in the mean responses of male and female students on the level of word processing skills acquired by secondary school students in Nsukka local government area of Enugu State.

Ho_2: There is no significant difference in the mean responses of male and female students on the level of graphic application skills acquired by secondary school students in Nsukka local government area of Enugu State.

Ho_3: There is no significant difference in the mean responses of urban and rural students on the level of Internet skills acquired by secondary school students in Nsukka local government area of Enugu State.

Research Methodology

The study adopted survey research design. This design is considered appropriate for this study because it permits the collection of data from the respondents on the level of computer appreciation skills acquired by secondary school students using structured questionnaire. The study was conducted in Nsukka Local Government Area of Enugu State. Nsukka local government have 30 secondary schools consisting of male and female secondary schools as well as mixed schools with only 13 schools out of the 30 secondary schools offering Computer Studies (Post Primary School Management Board, 2013). The population for the study consists of 2,229 senior secondary school students (SSS3) drawn from the 13 secondary schools in Nsukka Local Government Area that offers Computer studies. Purposive sampling technique was used to select 3 secondary schools from the 13 secondary schools that offer computer studies in their curriculum. Purposive sample is used when a researcher relies on his or her own judgment when choosing members of population to participate in the study (Black, 2010; Saunders, Lewis & Thornhill, 2012). These three secondary schools are: St. Theresa's College, Nsukka with 436 students, Community Secondary School Obukpa having 179 students and Urban Girls Secondary School, Nsukka with 228 SSS3 students. Simple random sampling through the use of balloting was then used in selecting the respondents from the three secondary schools which ensured that each participant have equal chances of being selected. The sample size was statistically determined using Taro Yamane formula for finite population, hence, simple random sampling was use to select the 272 participants used for the study (Yamane, 1967).

The instrument for data collection was a structured questionnaire titled Computer Appreciation Skills Questionnaire (CASQ). The questionnaire was made up of 18 items divided into three sections in line with the three research questions. The questionnaire was divided into two parts. Part one was the demographic variables of participants while part two was sub-divided into 3 sections: A, B and C each corresponding to the three research questions on the level of computer appreciation skills acquired by SS3 students. Section "A" contains 8 items which sought to elicit information on the word processing skills acquired by SS3 students. Section "B" was made of 5 items which sought to elicit information on graphic skills acquired by SS3 students while section "C" was made of 5 items which sought to elicit information on the Internet skills acquired by SS3. The questionnaire items was designed based on modified 4-point Likert scale rating of Very High Skilled, Highly Skilled, Moderately Skilled and Poorly Skilled with assigned weights of 4, 3, 2, and 1 respectively.

The instrument was face-validated by three research experts. The reliability of the instrument was established using Split-half reliability method. A pilot study of 20 SS3 students from Girls Secondary School Umulokpa in Uzo-Uwani Local Government Area of Enugu State which is outside the study area was used to establish the reliability of the instrument. Guttman split half coefficient (r) obtained for the various clusters were.72,.82 and.80 respectively thus indicating that the reliability index is high and is therefore consistent in measuring what it intends to measure. To ensure high return rate of the instrument three research assistants were engaged in the distribution and collection of questionnaire. The data collected was analyzed using mean and standard deviation for research questions while t-test was used to analyze the three null hypotheses at 0.05 level of significance. With respect to the hypotheses, any item where the p-value is greater than 0.05, the hypothesis of no significant difference was upheld at 0.05 level of significance and at 271 degrees of freedom; but where the p-value is less than 0.05, the hypothesis of no significant difference was rejected.

RESULT

First, the researchers examined the demographic information of participants for the study. Table 2 presents the demographic information of the participating students in secondary schools in Nsukka local government area of Enugu State. It can be seen that participants from STC constitutes 51% while 27% and 21% of participants were from Urban Girls' and CSS Obukpa respectively. With respect to gender, male students were 60% while female lecturers were 40% indicating that there are more male students in these schools than female.

Similarly, 26% of the participants (68%) were less than 14years of age, 41% were found to be between 15-18years of age while 32% of the participants were above 20years This also indicated that majority of the participants were mostly young adolescents. With regard to the school location and access to school learning, 73% of the participants reside in urban areas while 27% reside and school in rural areas.

Table 1. Descriptive statistics on the demographic information of the participating lecturers

Class		N	Percentage (%)
Name of school	St. Theresa's College, Nsukka	140	51
	Community Secondary School, Obukpa	74	27
	Urban Girls Secondary School, Nsukka	58	21
Gender	Male	164	60
	Female	108	40
	No	110	32
Age	<14yrs	74	26
	15-18yrs	112	41
	Above 20yrs	86	32
Location of school	Urban	198	73
	Rural	74	27

Table 2. Mean and standard deviations ratings of word processing skills acquired by secondary school students

S/N	Items	$\bar{x}$	SD	Rem.	t	df	Sig.	Rem.
1	Create, edit, and format document	2.92	0.95	HS	.650	270	.516	NS
2	Cut and paste document	2.95	0.92	HS	-.970	270	.333	NS
3	Copy and paste document	2.89	0.93	HS	-2.21	270	.028	NS
4	Maximize and minimize windows	3.01	0.91	HS	-2.19	270	.029	NS
5	Locate and use function keys	2.67	1.00	HS	.780	270	.436	NS
6	Highlight a text	3.01	0.96	HS	-2.18	270	.030	NS
7	Create and delete files from folder	2.47	1.03	MS	.201	270	.841	NS
8	Save document into disk or other devices	2.55	1.05	HS	-.586	270	.558	NS
	Grand mean	**2.81**	**0.97**	HS	.750	270	.350	NS

Key: $\bar{x}$ = mean, SD = standard deviation, rem = remark, t = t-test, df = degree of freedom, sig. = p-value, HS = highly skilled, MS = moderately skilled, NS= not significant, S = significant

Similarly, results presented in Table 2 show the level of word processing skills acquired by Secondary School Students (SSS3) in Enugu State. Table 2 indicated that students' acquire word processing skills highly in the areas of creating, editing and formatting of documents; maximizing and minimizing of windows; cutting, copying and pasting of document; locating and using of function keys; highlighting a text and saving documents into disk or other devices with mean ratings ranging from 2.55- 3.01 and the standard deviation ranging from 0.91- 1.05, thus, indicating that the respondents opinions are not far from each other. The cluster mean and standard deviation of 2.81 and 0.97 shows that the acquisition of word processing skills by Secondary School Students (SSS3) is at a high level.

Similarly, the researchers examined whether there is a significant difference in the mean responses of male and female students in Nsukka Local Government Area of Enugu State on the level of word processing skills acquired by secondary school students. The t-test analysis on Table 2 shows the results of item analysis as well as the cluster mean of the items. The results indicated that there is no significant difference in the mean responses of male and female students in Nsukka Local Government Area of Enugu State on the level of word processing skills acquired by SSS3 students. Since the p-value of 0.35 is greater than 0.05; the null hypothesis was therefore accepted as postulated not to have any significant difference at 0.05 level of significance.

Table 3. Mean ratings and standard deviations of graphic skills acquired by senior secondary school students

S/N	Items	$\bar{x}$	SD	Rem	T	df	Sig.	Rem.
9	Identify graphic packages	2.69	1.03	HS	.133	270	.894	NS
10	Identify features of the paint environment	2.61	1.08	HS	-.44	270	.657	NS
11	Draw simple object	1.88	0.91	HS	-.27	270	.784	NS
12	Make simple design and colour simple objects	1.99	0.96	MS	1.59	270	.112	NS
13	Modify a drawing by changing the size, colour and placement	1.99	0.95	MS	1.47	270	.141	NS
	Grand mean	**2.23**	**0.99**	**MS**	**0.23**	**270**	**0.31**	**NS**

Key: $\bar{x}$ = mean, SD = standard deviation, rem = remark, t = t-test, df = degree of freedom, sig. = p-value, HS = highly skilled, MS = moderately skilled, NS= not significant, s= significant

Furthermore, the researchers examined the level of graphic application skills acquired by Secondary School Students (SSS3). Result in Table 3 indicates that students are moderately skilled with regard to items 11, 12 and 13 with mean ratings of 1.88, 1.99 and 1.99 respectively. The computed cluster means of 2.23 and standard deviation of 0.99 shows that the students' level of acquisition of computer graphic skills is moderate low. The standard deviation rating which ranges from.91-1.08 shows that the respondents' opinions were not far from each other.

Furthermore, the t-test analysis presented in Table 3 shows that the p-value of 0.31 is greater than 0.05; null hypothesis was therefore accepted as postulated at 0.05 level of significance not to have any significant difference in the mean responses of male and female secondary school students in Nsukka Local Government Area of Enugu State on the level of graphic application skills acquired.

Finally the research examined the level of Internet skills acquired by Secondary School Students in the study area. The participants were requested to respond to the instrument and their responses were presented in Table 4. The results indicates that students' level of performance on Internet skills is high with items 14, 17 and 18 having mean ratings of 2.81, 2.90 and 2.93 and standard deviations of 0.99, 0.97 and 0.97 respectively. The item statements are: identify internet browsers, write valid e-mail address and recognize and name icons in the internet environment. Meanwhile, the overall cluster mean of 2.64 with a standard deviation of 0.98 indicates that the students' level of acquisition of Internet skills is high.

Also, the t-test analysis presented in Table 4 showed that the cluster means, t-calculated (0.29), degree of freedom (270) and the p-value (.18) to be greater than 0.05, thus, the null hypothesis was therefore accepted as postulated not to have any

Table 4. Mean ratings of internet skills acquired by senior secondary school students (SSS3)

S/N	Items	$\bar{x}$	SD	Rem	t	df	p-value	Rem.
14	Identify internet browsers	2.81	0.99	HS	2.253	271	.025	NS
15	Use internet search engines to obtain information and upload files	2.28	0.97	MS	-2.80	271	.005	S
16	Ability to create e-mail accounts and download files	2.26	0.98	MS	1.221	271	.223	NS
17	Enter valid e-mail address and password	2.90	0.97	HS	-1.60	271	.109	NS
18	Recognize and name icons in the Internet environment	2.93	0.97	HS	.618	271	.537	NS
	Grand mean	**2.64**	**0.98**	HS	.29	271	.18	NS

Key: $\bar{x}$ = mean, SD = standard deviation, rem = remark, t = t-test, df = degree of freedom, sig. = p-value, HS = highly skilled, MS = moderately skilled, NS= not significant, s = significant

significant difference at 0.05 level of significance in the mean responses of male and female secondary school students (SSS3) in Nsukka Local Government Area of Enugu State on the level of Internet skills acquired.

DISCUSSION

The findings of the study shows that SS3 students acquire an appreciable level of word processing skills such as ability to create, edit and format of documents; ability to use the keyboard expertly to retrieve and save documents into different drives and directories thus indicating the high level of skills possessed by SS3 in readiness for life after school. This finding is in line with Angib (2005) who noted that education is complete and meaningful only when it equips the citizens with the knowledge, skills and potentials necessary to live a fulfilling life and function effectively in the society.

In terms of graphic application skills acquired by secondary school students, the findings of the study show that students' acquisition of graphic skills are moderate in the use of graphic packages as seen in their ability to use software packages like Paint Artist, Adobe PageMaker, Microsoft Publisher, CorelDraw, among others, ability to use tools to make designs and colour objects, modify a drawing by ungrouping, changing the size, colour and regrouping. Furthermore, students seem to perform at a high level in some other operations such as identifying graphic packages and identifying their features which are more of recounting facts than the psychomotor skills that the students performs moderately. This finding supports the views of Federal Ministry of Education (2007) that students' ability to apply computer graphic skills such as CorelDraw to make simple designs like complimentary cards, business cards, wedding card, birthday, season greetings cards, logo etc will serve as sources of income for sustainability.

On the level of Internet skills acquired by secondary school students, the findings of this study shows that Internet skills are highly acquired by students in the area of using search engines to download and upload information, create e-mail account, sending and receiving mails as well as engaging social networking sites like google scholar, wikis, whatsapp, social chart rooms, among others for collaborative studies. This finding is in line with Post Primary School Management Board (2013) assertion that SSS3 students should be able to use Internet facilities to create e-mail account and link up with educational network groups (e.g. school, Net, e-school etc) to enhance their knowledge base.

IMPLICATIONS

The findings of this study have a number of implications in terms of quality of education offered to our secondary school students in the present knowledge driven economy. The result showed that students' acquisition of word processing skills is high with students in urban areas performing much higher than those in the rural areas. This is an indication that students can type their assignments, letters, reports, minutes of meetings etc performing various task of creating, editing, formatting, saving and printing of documents are at high level of performances. Furthermore, the low level of acquisition of computer graphic skills by students' calls for programme review in terms of number of hours that students should be engaged in drills and practice, minimum number of computer systems available in every school and ICT teachers available in every school.

The result also revealed that students' acquisition and use of Internet skills is at a very high level. This implies that students can connect with friends and families in distant land, communicate and collaborate with people of common interest and finding solutions to school work through electronic mails, yahoo messenger and search engines. Result also revealed that students in the urban schools acquire more computer skills in areas like Internet, spreadsheet, data processing, word processing, installation and troubleshooting than their counterparts domicile in rural communities. The implication is that students in the urban schools have better access to ICT resources like computers, internet facilities, printers, as well as well qualifies teachers who prefer to live in urban centres than those living in the rural areas thus creating room for high rural-urban migration for students and teachers alike.

CONCLUSION

Application of computer studies and general ICTs at all levels of schooling especially at primary and secondary schooling is seen to play a vital role in manpower development required for sustainable development of any nation. Secondary school students (SSS) are seen to be curious to learn especially when learning involves working with computers. Students' curiosity to learn should be encouraged by all stakeholders of education since skill development demand adequate preparation, proper supervision and monitoring of teachers to ensure optimum service delivery. Adequate computer skill training and preparation has obvious implications for students, educational system, parents and the society at large. For example, the students in the rural communities are seen to be lagging behind their counterparts in the area of Internet access and usage, inadequate computer teachers and most importantly electricity. Schools in the rural communities' school should be equipped with general ICT facilities, access to electricity and provision of well trained ICT teachers to help bridge the gap. It is therefore the duty of all stakeholders within and outside the educational sector to evolve appropriate means of ensuring functional education at all levels of education in Nigeria.

RECOMMENDATIONS

Based on the findings of the study, the following recommendations can be suggested:

1. The ministry of education should equip all secondary schools especially those in the rural areas with adequate ICT resources needed to implement functional education in Nigeria.
2. Parents should expedite action and provide their wards access to computers and Internet connectivity at home so as to complement the training giving to them at school.
3. Remuneration of teachers in the field of computer and ICTs should be reviewed to be at par with their counterparts at the private sectors so as to attract them to the teaching industry.
4. Donor agencies, international organizations, NGOs and good spirited individuals should be involved in developing and supporting strategies that will enhance ICT integration in schools through donation of one dollar laptops to schools, retraining programs for teachers in area of ICT and counterpart funding.

REFERENCES

Achuonye, K. (2003). Towards an effective integration of the Information Technology (IT) into Teacher Education. *Eastern COEASU Journal of Teacher Education*, *1*(1), 83–87.

Aigbovbiosa, O. I. (2000). Reforms in physics education for sustainable national development. *Mosogar Journal of Science Education*, *2*(10), 70–71.

Ajayi, G. O. (2000). *Information Age: Virtual Technology Opportunities and Challenges in Africa*. A paper presented at UNESCO National Workshop on the Virtual Universities/Laboratory Project at OAU, Ile-Ife, Nigeria.

Alabi, A. T. (2001). Availability of information technology equipment for instructional purpose in Ilorin metropolis secondary schools. *Nigerian Journal of Educational Research and Evaluation*, *3*(2), 16–19.

Allen, J., & Van der Velden, R. (2012). Skills for the 21st Century: Implications for Education. ROA-RM- 2012/11. Maastricht: ROA.

Angib, M. U. (2005). The family as a supportive Agent Computer Education of a Nigerian Child. *A Journal of the New Nigerian Educators, 1*(1), 86-94.

Babalola, J. B. (2007). *Access, equity and quality in higher education*. NAEAP.

Bakpo, F. S. (2005). *Introduction to Computer Science and Programming in QBASIC*. Enugu: Magnetic Business Enterprises.

Black, K. (2010). *Business Statistics: Contemporary Decision Making* (6th ed.). John Wiley & Sons.

Candy, M. (2000). End of Millennium- Information Age. Economy, Society and Culture, 1(3), 10-13.

Eisenberg, M. B., & Johnson, D. (1996). *Computer skills for information problem-solving: learning and teaching technology in context*. ERIC Digest.

Ejionueme, L. K. (2007). *Strategies for involving the Private Sector in the Management of Public Secondary Schools in Nigeria*. A paper presented at Institute of Education Conference on Optimization of Service Delivery in the Education Sector: Issues and Strategies, Nsukka, Nigeria.

Federal Ministry of Education. (2004). *National Policy on Education* (4th ed.). Lagos: NERDC Press.

Federal Ministry of Education. (2007). *9-Year Basic Education Curriculum: Nigerian Education Research and Development Council.* Abuja: UBEC Press.

Gbadeyan, J. A. (2005). Computer and research activities. In H. A. Saliu, J. A. Oyebanji, & A. Jimoh (Eds.), *Basic issues in research methodology. Ilorin: T.A. Olayeri Printers & Publisher.*

Grass Root Design. (2004). *Introduction to Computer-A Beginners Guide*. London: Chapman and Hall.

Howland, J. L. (2013). *Facts101: Textbook Key Facts*. Contents Technologies Inc. Retrieved 2015-10-03 from https://en.wikipedia.org/wiki/Skill

King, K. (2005). Re-targeting schools, skills and jobs in Kenya: Quantity, quality and outcomes. *International Journal of Educational Development*, *25*(4), 423–435. doi:10.1016/j.ijedudev.2005.04.005

King, K., & Palmer, R. (2007). *Skills Development and Poverty Reduction: A state of the art review.* European Training Foundation. Accessed on 23/11/2016 from www.etf.europa.eu/pubmgmt.nsf/(getAttachment)/.../$File/NOTE79TKHJ.pdf

McLuhan, M., & Bruce, R. P. (1989). *The Global Village: Transformations in World Life and Media in the 21st Century*. New York: Oxford University Press.

NCREL & Metiri Group. (2003). *enGauge 21st century skills: Literacy in the digital age*. Authors.

Obi, C. A. (2005). *Methodology in Business Education*. Enugu: Oktek Publishers Nig. In Association with De – Verge Agencies Ltd.

Offorma, G. C. (2002). *Curriculum Theory and Planning*. Enugu: Family Circle Publications.

Okporie, B. U. K. (2014). Evaluation of the operational activities in the construction industry and the way forward for sustainable development.*Proceedings of International conference on research, innovations and sustainable development*, *6*(6), 215-219.

Olelewe, C. J., & Nzeadibe, C. A. (2015). Challenges facing the Availability and Utilization of ICT resources in Post Primary Schools in Nsukka Educational Zone of Enugu State, Nigeria. In *Information Communication Technology (ICT) Integration to Educational Curricula: A New Direction for Africa*. Available at http://www.booktopia.com.au

Omeregie, N. (2005). *Re-packaging secondary education in Nigeria for great and dynamic economy.* A paper presented at the 2nd Annual National Conference of Association for Encouraging Qualitative Education in Nigeria, Ekpoma, Nigeria.

Osinem, E. C., & Nwoji, U. C. (2005). *Students Industrial Work Experience in Nigeria: Concepts, Principles and Practice.* Enugu: Cheston Agency Ltd.

Osuala, E. C. (2004). *Principles and Methods of Business and computer Education.* Enugu: Cheston Agency Ltd.

Oyaziwo, A., Philipa, O. I., & Imonikhe, J. S. (2012). Increasing access to university education in Nigeria: present challenges and suggestions for the future.*The African Symposium: An Online Journal of the African Educational Research Network, 3*(12), 323-340.

Pacific Policy Research Center. (2010). *21st Century Skills for Students and Teachers.* Honolulu, HI: Kamehameha Schools, Research & Evaluation Division.

Pollak, D. (2009). *Neurodiversity in higher education: Positive responses to specific learning differences.* New York: Wiley. doi:10.1002/9780470742259

Post Primary School Management Board. (2013). *Statistical data of Secondary Schools in Nsukka Educational Zone of Enugu State.* Academic Press.

Saunders, M., Lewis, P., & Thornhill, A. (2012). *Research Methods for Business Students* (6th ed.). Pearson Education Limited.

Shapiro, J. J., & Hughes, S. K. (1996). Information literacy as a liberal art: Enlightenment proposal for a new curriculum. *EDUCOM Review, 31*(2), 31-35. Available at: http://www.educause.edu

Sheldon, A. (1999). *The cultural impact of Computer Technology: Contents of Curriculum Unit 99.07.07.* Yale-New Haven Teachers Institute. Accessed on 22/11/2016 from http://teachersinstitute.yale.edu/curriculum/units/1999/7/99.07.07.x.html#h

Smart-Board Technology Integration in Teaching. (2011). *AMCIS 2011 Proceedings-All Submissions. Paper 18.* Accessed on 1st November, 2016 from http://aisel.aisnet.org/amcis2011_submissions/18

Smith, P., Cowie, H., & Blades, M. (2003). *Understanding children's development* (4th ed.). Blackwell Publishing.

Taggart, B. (2007). *The effective provision of pre-school education (EPPE) project: Intensive case studies of practice across the English foundation stage.* Keynote address at the CECDE Conference Vision into Practice, Dublin, Ireland.

Torane, S. (2010). *Secondary Education Definition. Missouri Department of Elementary and Secondary Education*. Retrieved on September 25, 2010 at http:// www.2010secondaryeducationdefinition.com

Trade Economics/National Bureau of Statistics. (2012). *Nigeria Population*. Available: http://www.tradingeconomics.com

UN Department of Economics and Social Affairs (UNDESA). (2008). *Sustainable Development*. Retrieved from www.un.org/esa/sustdev/sdissues.htm

United States Embassy in Nigeria. (2012). *Nigeria Education Fact Sheet*. Author.

Voogt, J., & Pareja, R. N. (2010). *21st Century Skills: Discussienota*. Enschede: Universiteit Twente.

Weaver, G. C. (2000). An examination of the National Educational Longitudinal Study (NELS: 88) database to probe the correlation between computer use in school and improvement in test scores. *Journal of Science Education and Technology*, *9*(1), 121–133. doi:10.1023/A:1009457603800

Wokocha, S. (2013). *ICT as a 21st century standard for Nigeria teachers*. University of Lagos: *West African Conference and Exhibition*.

World Bank. (2002). Constructing Knowledge Societies: New Challenges for tertiary education. Washington, DC: Academic Press.

Chapter 11
Conceptual Design and Usability Evaluation of a Grid Utility SECOnD Portal for Under-Resourced Business Enterprises in Sub-Saharan Africa

Babafemi Odusote
Covenant University, Nigeria

ABSTRACT

Conventional grid-enabled portal designs have been primarily based on the usual traditional requirements specifications such as security requirements, grid resource requirements, job management requirements and the likes. However, the pay-as-you-use service provisioning model of utility computing platforms need to consider and satisfy certain salient requirements considered relevant in a bid to achieve better conceptual designs of Service-based Electronic Commerce On-Demand (SECOnD) that are well-usable for grid utility platforms such as the Grid-based Utility Infrastructure for Small, Micro, and Medium Enterprises (SMME) Enabling Technology (GUISET). The work aspect profiled in paper entails the conceptual design and usability evaluation of a SECOnD portal which serves as the uniform channel through which targeted under-resourced SMMEs can inexpensively access and use e-Commerce technologies and tools without owning them. The outcome of the evaluation experiment revealed that the service portal designed largely achieve the usability expectations of it users.

DOI: 10.4018/978-1-5225-2565-3.ch011

INTRODUCTION

A grid amongst many ways can be defined as a collection of heterogeneous distributed interconnected computing and data resources that provides large scale virtual resources with an appropriate single user interface (Foster & Kesselman, 1999). The utility grid computing paradigm is a variation of the conventional grid computing model that involves the provisioning & utilization of resources and services solely on per user request basis (Rappa, 2004). It is essentially based on the *pay-as-you-use service* provisioning business model.

In a typical grid environment, grid portals are designed and developed to provide uniform access or single point of entry to various underlying grid resources and services within a virtual community (Russell, Novotny & Oliver, 2006). They act as gateways between various grid users and a range of different grid resources and services (Russell, Novotny & Oliver, 2006).

Grid-enabled portals (GeP) - portals based on grid technology, build upon the familiar Web portal model to offer virtual organizations (VO) or communities of users a single point of access to computational resources such as clusters, data servers, applications, scientific tools, and computing services (Russell et al., 2006; Yanli, Jian, Minglu, Lei, 2006). They are key technologies that have received great attention among developers and programmers due to their ease in development, richness in functionality, pluggable architecture and customization of interfaces (Russell et al., 2006; Yanli et al., 2006).

Grid-enabled portals are developed using *portlets* (Yanli et al., 2006). Portlets can be thought of as a miniature web application that is running inside a portal page alongside a number of similar entities. They are user-facing, multi-step, interactive modules that can be plugged into a portal application (Yanli et al., 2006). Portlets rely on the overall portal framework to access user profile information, participate in a presentation interface, communicate with other portlets to access remote contents, lookup credentials, and store persistent data (Yanli et al., 2006; Akram, 2005). Figure 1 shows portals composed of various portlets.

Portal frameworks are development platform for portal development. They are design structures that contain various modules, methods and software features used for developing specific portal (Russell, Novotny & Oliver, 2006; Akram, 2005). With the popularity of portals today, there are many portal frameworks that are available as open source and the list of these open source frameworks is all the time increasing. A typical grid-enabled portal has the following capabilities Russell, Novotny & Oliver (2006): secured registration of users; accessibility of various users to a range of underlying grid services and resources; provision of personalization, single sign-on (SSO), aggregation and customization; robust application integration; security; redundancy, failover and load balancing etc.

Portlets are used in portal development as self-contained pluggable user interface components to encapsulate one or more applications or services that can be aggregated and transferred as information to a presentation layer of a portal system (Yanli et al., 2006; Akram, 2005).

Hitherto, conventional grid-enabled portal designs have been primarily based on the usual traditional requirements specifications (Russell, Novotny & Oliver, 2006; Akram, 2005). However, the pay-as-you-use service provisioning model of utility computing platforms necessitate the need to satisfy certain salient additional requirement specifications considered relevant in a bid to achieve better conceptual designs of grid-enabled portals that are well-usable for utility computing platforms (Eilam, 2004; Phillip, 2004).

This paper presents a conceptual design and usability evaluation of a grid utility SECOnD portal that caters not only for the traditional requirements but also the relevant additional requirements for utility computing contexts.

In addition, it presents formal guidelines for the conceptual design of grid utility service portals that can help to foster the practice of design and development of grid utility SECOnD portals.

Figure 1. Portals composed of various portlets

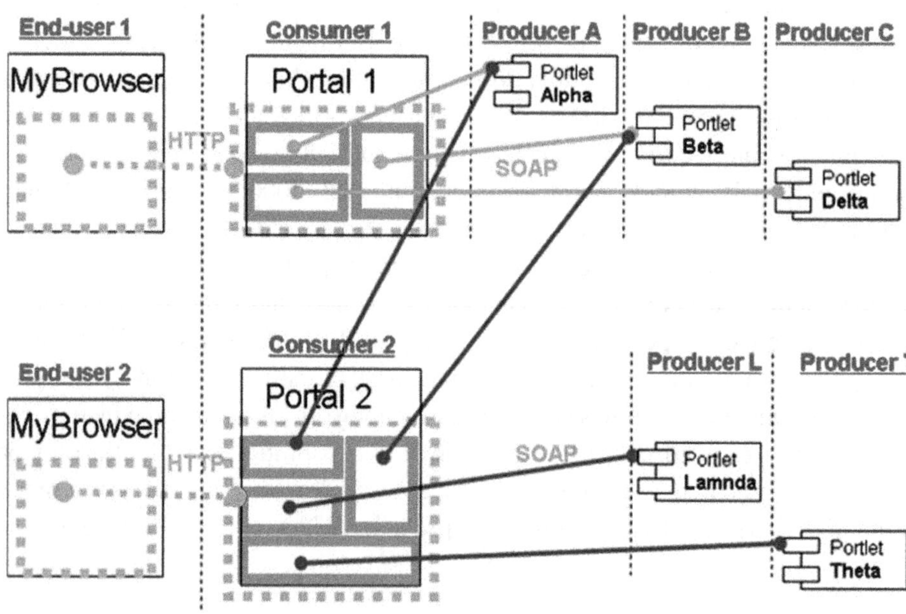

Based on the conceptual design, a prototype SECOnD portal for GUISET was developed. GUISET is aimed at technologically enabling the business activities of SMMEs in Africa by facilitating an affordable access to relevant technologies on a *pay-as-you go* basis (Adigun, 2008; Odusote & Adigun, 2014). The GUISET portal is equipped with a set of features to address these additional requirements.

A usability evaluation controlled experiment was performed using a group of users to evaluate the perceived important additional requirements using a post-task questionnaire based on the Technology Acceptance Model. (Alharbi & Drew, 2014). The evaluation results revealed that these set of features in the GUISET portal were perceived by the users to fulfill some of the additional requirements for UC.

The rest of the paper is organized as follows. Section 2 presents the review of some of the related works on Grid portals and portlet technology. Section 3 shows the conceptual design of the Grid-enabled portal. Section 4 presents the prototype of the Grid-enabled portal. In Section 5, the usability evaluation experiment was briefly described. The outcome of the evaluation experiment revealed that the service portal achieved the usability expectations of its users. The research work was summarized and conclusions were made in Section 6.

BACKGROUND AND RELATED WORKS

Many of the early versions of grid portals are non portlets-based, for example, the Astrophysics Simulation Collaboratory (ASC) portal, UNICORE, etc. (Russell et al., 2006; Yanli et al., 2006]. However, the development of portlets-based portals has brought many benefits to both end-users and developers, and this is getting more recognition. An investigation of a set of related works done in this research work revealed that, a combination of Grid technologies and portal technologies has been substantially achieved in order to make grid-enabled portals more flexible and to improve their usability (Yanli, 2006).

From a user's perspective, a Grid based portal system usually should provide the following functionalities or services:

1. Authentication Service

Any user that intends to log onto the portal using a web browser is first authenticated by means of a user-id and password. Once the user is authenticated, it is the job of the portal server to act as the user's proxy in most grid interactions, by obtaining a proxy certificate that it can use on behalf of the user. The standard approach is to have the user submit a proxy to a *MyProxy* server (Yang et al., 2006). The proxy can then be retrieved by the portal server from the MyProxy server and holds the proxy for the duration of the user's session. This process of making the user manage a key pair and submit proxy certificates to a MyProxy server is extremely unpopular with users.

2. Remote File Management

A central requirement for the portal is the access to file metadata directories and remote file archives. Simple tools for Grid file transfer protocol, (Grid FTP) are essential here, but many files are likely to be managed by a virtual data system, where data is cataloged and staged by back-end grid services.

3. Remote Job Submission and Monitoring

The ability to submit jobs to the grid infrastructure for execution and monitoring is a classic requirement for portal. Users with specific resources allocation want to be able to see job queues on those resources and consult scheduling assistants. They need to be able to keep track of job execution and understand when things fail by reading logs.

4. Access to Collaboration

Resource sharing is also vital within any VO. The ability to use real-time collaboration tools as well as asynchronous collaboration is emphasized here.

Grid systems have some common services or functionalities, and these services are independent from each other somewhat. Thus, one or more Grid services can be encapsulated into a grid portlet and enable portlets to interact with each other if needed (Yanli et al., 2006; Akram, 2005). An authentication portlet can be designed to authenticate users. A Grid FTP portlet can enable users to access remote files if they are authenticated users. A job submission portlet allows users to summit jobs and a resource monitoring portlet to enable users to check remote resources such as CPUs or memories. Collaboration portlets such as chatting portlet allows users to collaborate with each other. Grid users can choose the portlets they need to configure their own portals.

These traditional requirements/functionalities have primarily influenced the design of conventional grid-enabled portals (Russell et al., 2006; Akram, 2005). However, additional requirements that stem from computing models such as the pay-as-you-use service provisioning model of utility computing platforms present an opportunity for the conceptual design and development of effective & usable grid-enabled portals for such utility computing platforms (Eilam, 2004; Pagden, 2003; Phillip, 2004).

THE GRID UTILITY SECOND PORTAL DESIGN

Prior to this work, in Yanli (2006) a portlet-based Grid portal architecture was proposed for integrating existing technologies under a common interface. GT 3.2 (Globus Toolkits, 2014) was adopted as the underlying Grid system and a portal prototype was developed using JetSpeed 2 (www.portals.apache.org) which is Java Specification Request, JSR-168 compliant (www.developers.sun.com), as the portal framework. This work adapted this architecture to a standardized form by introducing the Web Services for Remote Portals, WSRP standard and specification (www.oasisopen.org). Consequent to the adapted framework, presented in Figure 2 is a conceptual layout of the SECOnD portal with respect to the various additional requirements which were catered for in the design of the portal for utility computing context.

The design lays out the various functional components and structures of the system and their relationship to one another. This design is profiled in detailed in (Odusote & Adigun, 2014). However, for better understanding, the various subsystem or modules of the service portal are highlighted in the following subsection with the respective use case designs. A logical design of the service portal is presented in fig 2.

Figure 2. The conceptual layout of the GUISET portal (Odusote & Adigun, 2014)

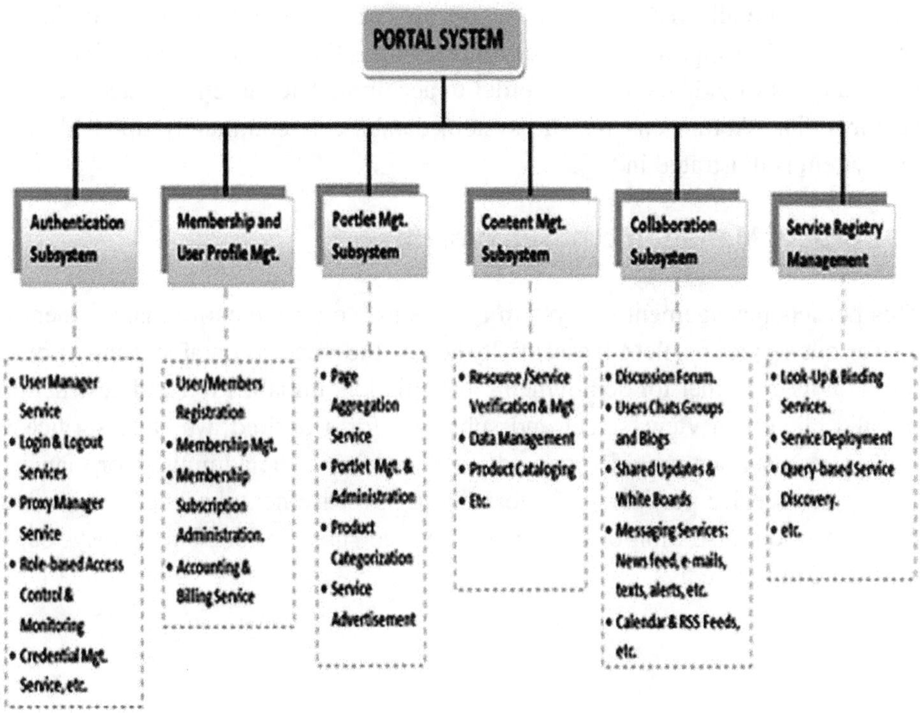

A. Authentication Subsystem

The authentication subsystem is designed to ensure that only valid users have access to the system based on stipulated policies such as Role-Based Access Control (RBAC). The administrator manages the various clients' access credentials alongside their logs and session. The key Actor here is the Administrator of the system who has the sole responsibility to administer the clients' access control by assigning role, manage clients/users access credentials and sessions logs. This Subsystem is illustrated in Figure 3.

B. Membership and User Profile Management Subsystem

This subsystem is responsible for the management of users' registration, and profile management. A valid client can setup a subscription request after viewing and selecting an available service. The administrator validates the client's registration and subscription request after ensuring that the appropriate requirements have been met. The Admin handles the overall management of clients' accounts subscription and also ensures the interests of the business clusters are protected. Each business cluster is registered on the GUISET infrastructure as a distinct business entity strictly based on the business domain or type of goods and services offered. Only SMMEs that qualify are allowed to belong to specific clusters, the unqualified applicants are channeled to appropriate clusters. However, as the business domain increases, more clusters would be created in order to accommodate more interested business ventures. The Admin ensures the system database is continually updated. This Subsystem is illustrated in Figure 4.

C. Portlets Management Subsystem

This portlets management subsystem is responsible for the entire management of the various service portlets. It essentially houses the portlets containers and a portlet relies on its container for deployment, instantiation, initialization and destruction. A valid client can view, select and subscribe for a desired available service as well as, as a service provider, can advertise services. The administrator manages the various service portlets – categorization, advertisement, backup, and ensures the proper administration of the client's service subscription. This Subsystem is illustrated in Figure 5.

D. Content Management Subsystem

This is responsible for the administration and management of the various portal contents. A valid client can view available services for consumption; and also upload & advertise services for use. The administrator verifies and manages these services for advertisement. This does the service cataloguing and data management. The Content Management Subsystem is illustrated in Figure 6.

E. Collaboration Subsystem

The collaboration subsystem is designed to enable coordinated interaction and collaboration amongst the various user communities. A valid user can join an existing user community based on interest or setup a new one, creating a user forum for similar business interest discussion. The Admin verifies and validates the Client's subscription requests, manages the clients' account subscriptions, and oversee the general interaction of all the business entities. This Subsystem is illustrated in Figure 7.

F. Service Registry Management Subsystem

There is a registry that serves as a service repository. A valid client can engage in a service look-up and binding based on the available service description in the registry. The administrator manages the service registry and periodically updates its content based on the client's requests and search. The services would be accessed through user service interfaces which are the interfaces of the various portlets which are designed to encapsulate the various services. This Subsystem is illustrated in Figure 8.

Sequence Diagram

Interaction among the various entities (actors) of the system are also captured with a sequence diagram. This is illustrated as follows: It begins with portlets registration by the portal administrator normally through a portal, Integrated Development Environment (IDE) and ends up with a portal being registered to a given portlet producer. The Figure 9 outlines the protocol involved in the sequence diagram.

First, an introductory description of the producer is obtained through the "getServiceDescription()" function. If registration is required then, the consumer must register with a producer before accessing any of the producer's portlets. Once registered, the consumer queries again the producer but now, a detailed description of the available portlets is returned. With all this information, the portal IDE creates a WSRP consumer. This WSRP consumer is within the portal realm.

Figure 3. Use case diagram of the authentication subsystem

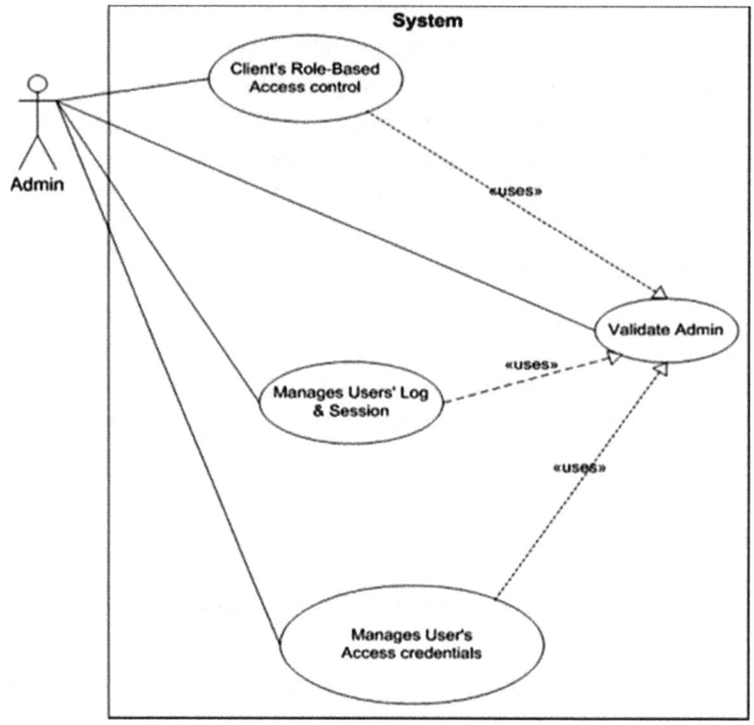

Once registered, the portal is ready to engage the portlet in conversation to deliver its service. This is achieved through a two-step protocol as shown in Figure 9. To begin with, the very first markup realizing the service is obtained through the "getMarkup()" function. The returned markup is aggregated to other markup that built up the portal page which is finally rendered to the end user. Whenever the user clicks on a link of the portlet markup, the portal receives the HTTP request which is in turn, forwarded to the portlet producer by means of the "performBlockingInteraction()" function till it finally reaches the portlet itself. As a result, the portlet can change its state. But no markup is returned to the consumer. This requires the consumer to issue a "getMarkup()" function to recover the eventually new markup associated with this new state.

THE PORTAL IMPLEMENTATION

A proof of concept design prototype of the Grid-Utility SEConD portal was built on the standardized portal development toolkit/framework, Liferay 5.2.3 portal toolkit

Figure 4. Use case diagram of the membership and user profile management

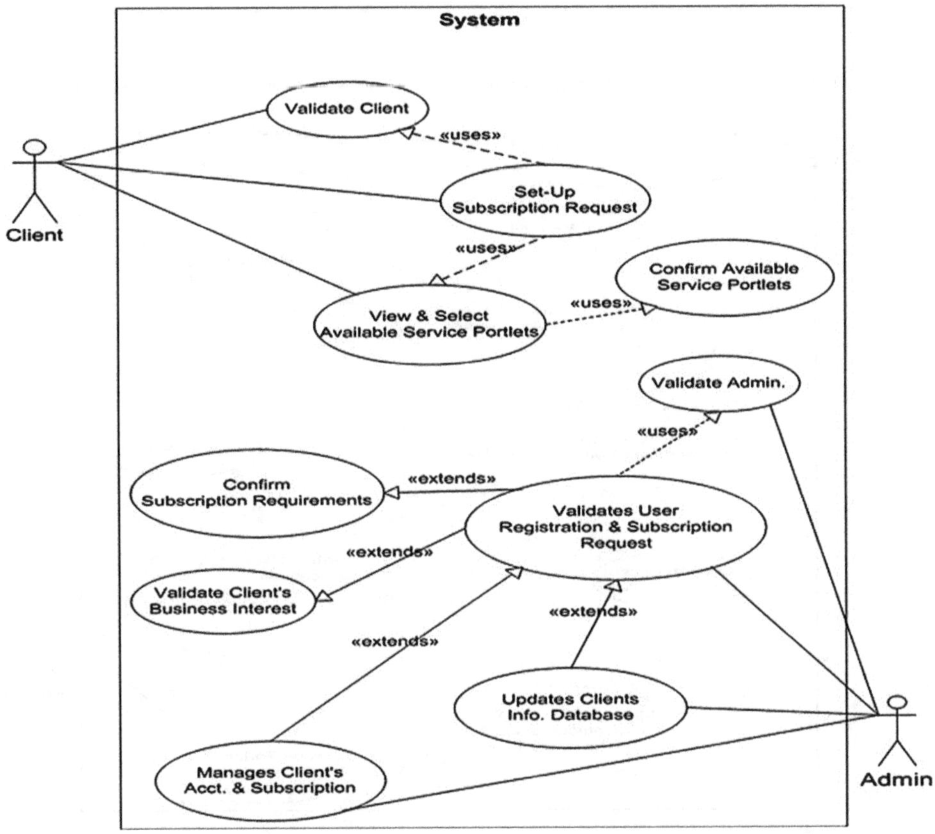

(www.liferay.com) bundled with Tomcat 6.0.18. This implementation details is reported in (Odusote & Adigun, 2014).

Figure 10 shows a snapshot of the portal's main configuration panel.

THE EVALUATION

The objective of the evaluation is primarily to determine if the salient requirements considered in literature to be relevant towards achieving better conceptual designs of the grid-enabled service portal for the grid utility platform helped to achieve the usability expectations of the various users. The usability evaluation of the grid-enabled portal was designed in a way that it can be achieved as effectively as possible. A set of relevant criteria for evaluating the usability of grid-enabled portals were identified and used as profiled in literature. A total of twenty-four

Figure 5. Use case diagram of the portlets management

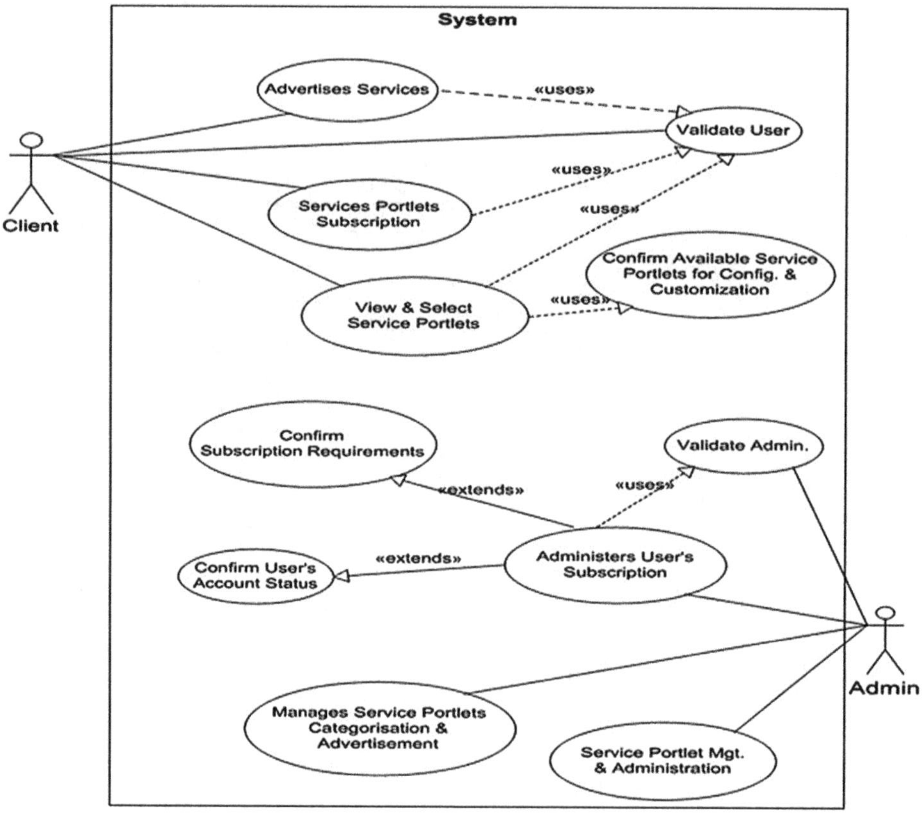

(24) randomly selected start-up business students-entrepreneurs from the Centre for Entrepreneurship Development, School of Business, Covenant University were taken through the evaluation process.

The selected participants were taken through a pre-experiment orientation, by being given various explanations on grid and utility computing concepts, the workings of grid-enabled portals and how to use them to achieve specified tasks. They were also taken through brief work-through training on what the various requirements and expectations are.

The purpose of conducting this brief was to make sure that participants have no difficulties to understand all steps of each task. This really helped to make the task description simpler and easily understandable. After a substantial level of understanding of each participant was ascertained, they were then assigned a set of tasks to perform on the portal before they filled the usability evaluation questionnaires.

The statistics on the background of the participants is presented in Table I. The result of the background test revealed that 75% of the participants have an above-average

Figure 6. Use case diagram of the content management

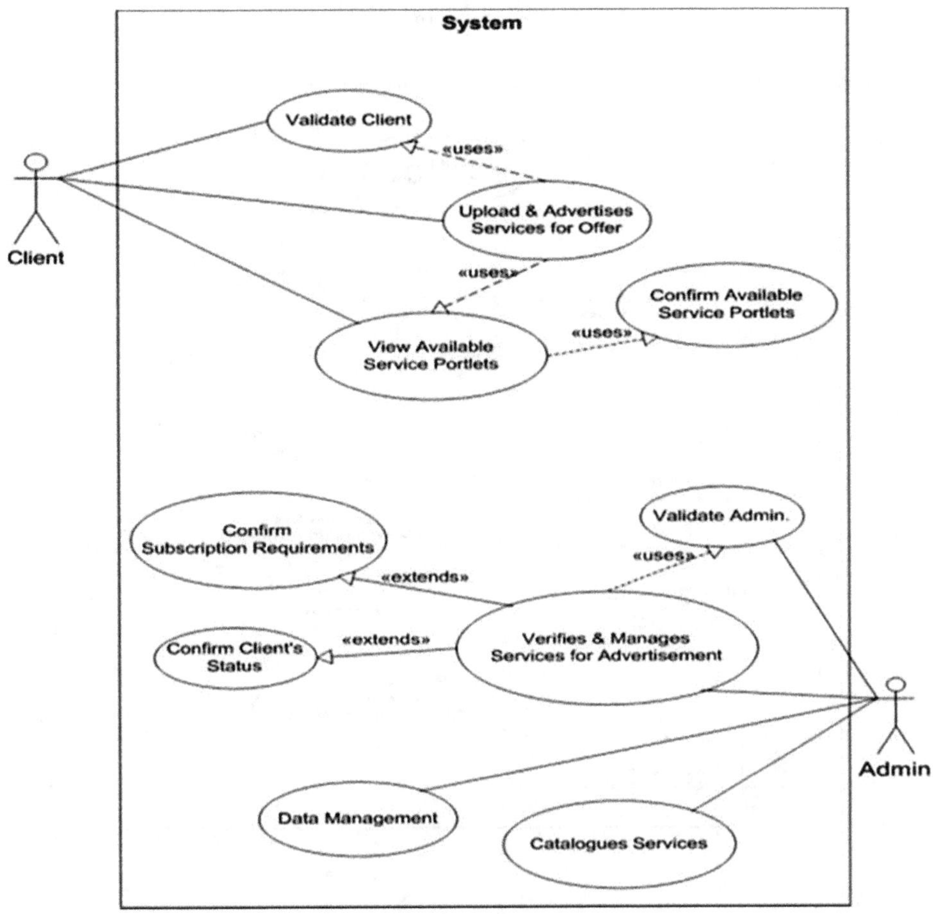

level of understanding of grid-based technologies and 66.66% of the participants has an above-average level of experience with the use of grid-enabled portals.

The evaluation questionnaire was designed by the author on the basis of findings from usability test and guidelines for usability evaluation of the web sites provided by IS&T Department, MIT. The selected criteria are numbered: 1. Simplicity; 2. Satisfaction; 3. Aesthetics; 4. Memorability; 5. Hypertext Structure; 6. Security; 7. Resourcefulness & Job Management; 8. Accounting; 9. Collaboration; 10. Messaging.

The overall response of the participants concerning the various identified & selected criteria on the questionnaires based on the additional requirements showed that an average value of 21.67% participants *strongly agreed* with the various criteria according to the questionnaire results. 28.33% participants *agreed* and the 21.67%

Figure 7. Use case diagram of the collaboration subsystem

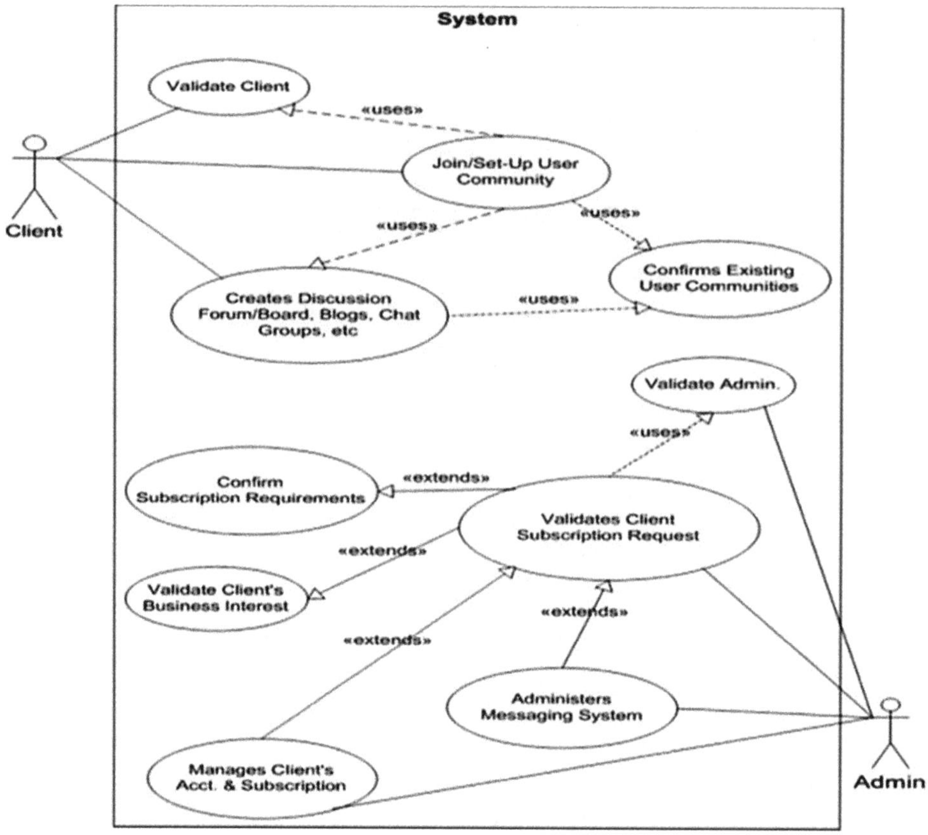

was *indifferent* in the questionnaires. Average 19.16% participants *disagreed* and 9.17% participants *strongly disagreed* with the various criteria according to the results.

The overall result is presented in Figure 11. The general observations from the additional analytical evaluation experiment as already reported in another paper (Odusote & Adigun, 2014) revealed that the service portal fulfilled the minimum requirements to be suitable for UC context, and as the research is taken further, it is hope that more of the requirements would be fulfilled.

SUMMARY AND CONCLUSION

Recent findings in practice of grid-enabled portal development for utility computing platforms necessitated the need for additional requirements to be fulfilled in order to realize usable grid-enabled portals for utility computing platforms. This work

Figure 8. Use case diagram of the service registry management

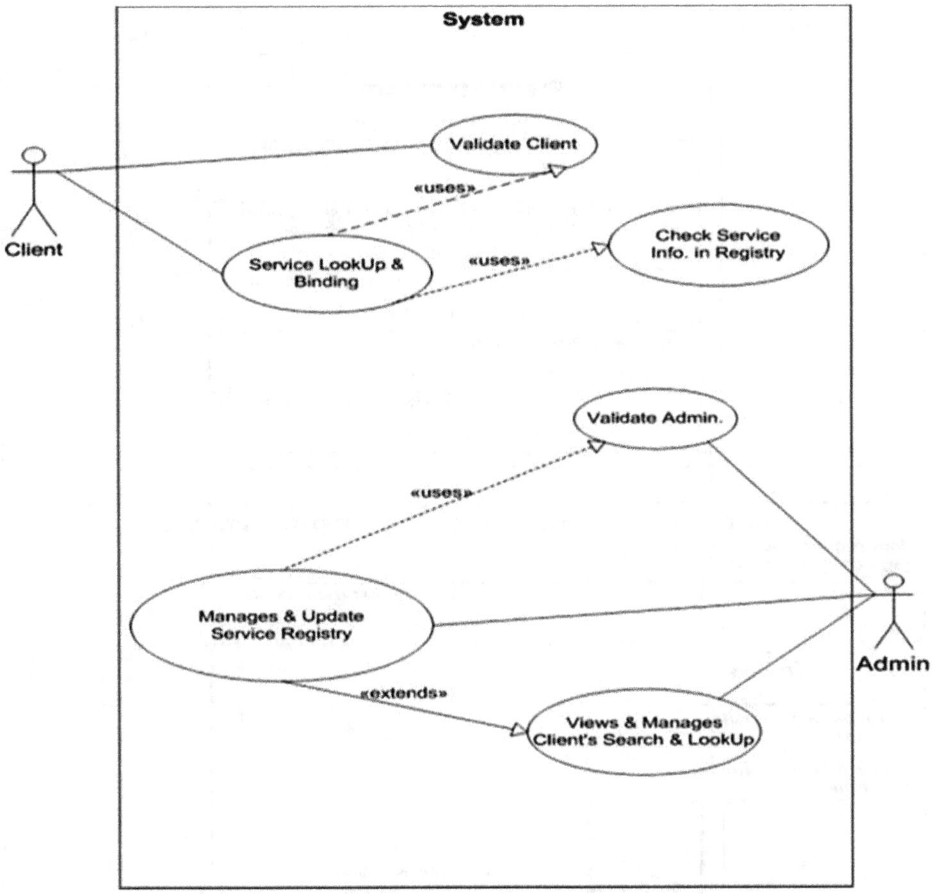

investigated and identified the minimum relevant additional requirements - the basis for the design of the SECOnD portal which serves as the uniform channel through which targeted under-resourced SMMEs can inexpensively access and use e-Commerce technologies and tools without owning them. The portal prototype is designed for GUISET initiative – a Grid Utility Computing platform that was conceptualized to technologically enable the business activities of SMMEs by facilitating an affordable access to relevant technologies on a pay-as-you-use basis. The SECOnD portal prototype serves as an interface meant to provide an enabling operating environment for every prospective utility service providers and customers willing to form or join any user business cluster or community.

Figure 9. Sequence Diagram (WSRP Protocol) (Oscar et al., 2007)

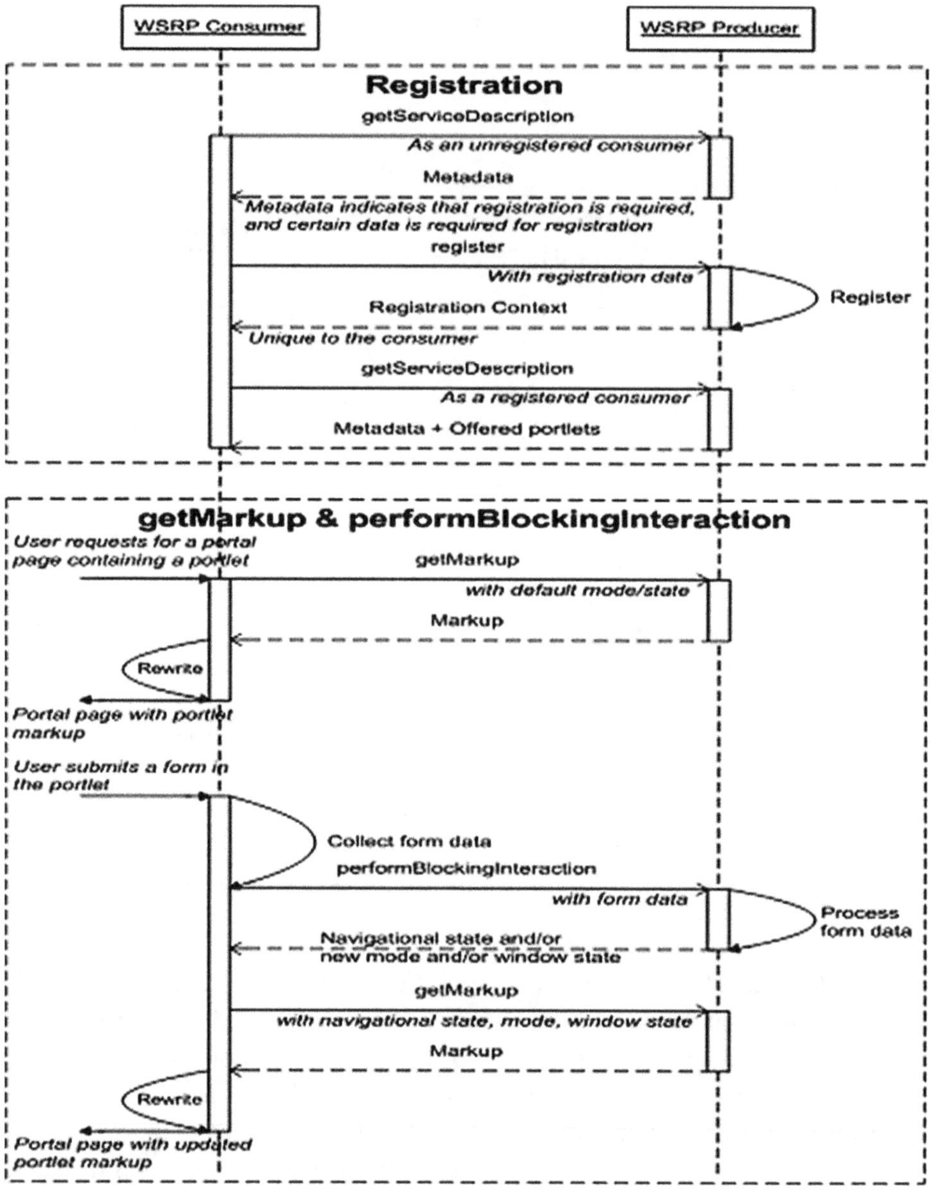

This study profiles template for the design and implementation of SECOnD portals as a uniform medium for affordable access to scarce and expensive computing capabilities, particularly among under-resourced SMMEs in Sub-saharan Africa.

Figure 10. The portal main configuration panel

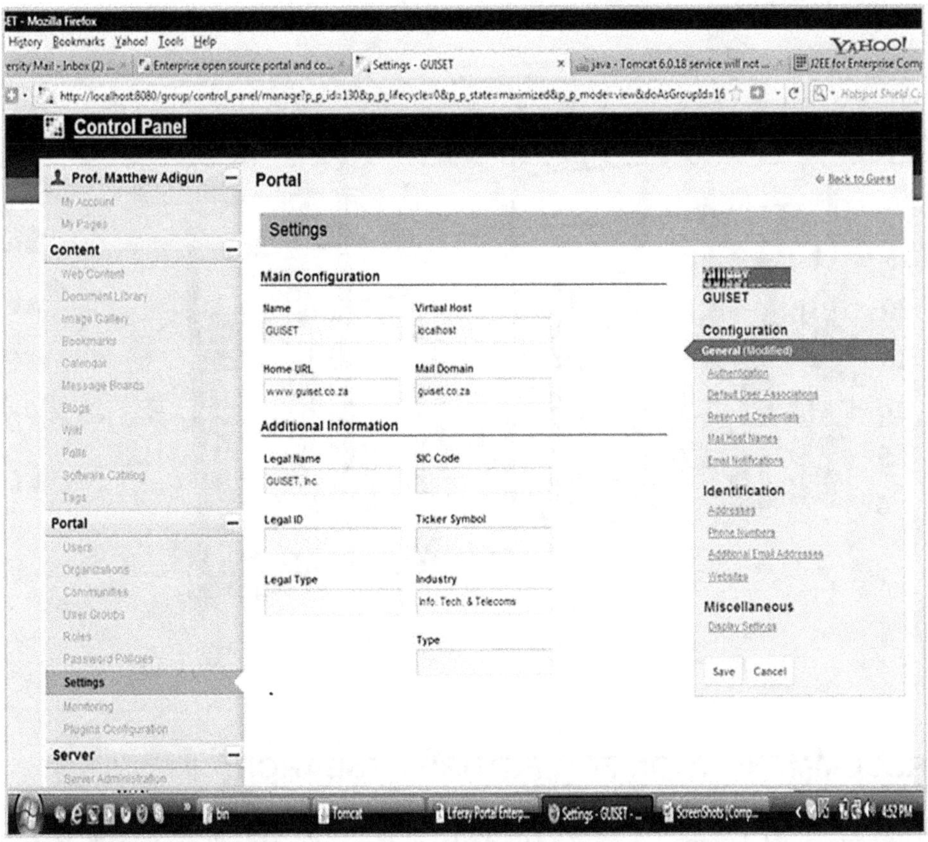

Table 1. The background of participants

Background of Participants	No. of Participants	Very High	High	Medium	Low	Poor
Level of understanding of Grid-based technologies	24	8.3%	25%	41.7%	16.7%	8.3%
Level of experience with use of Grid-enabled portals	24	8.3%	25%	33.3%	25%	8.3%

Figure 11. Graphical Representations of Participants' Responses
Key: A: Strongly Agree; B: Agree; C: No comments; D: Disagree; E: Strongly Disagree

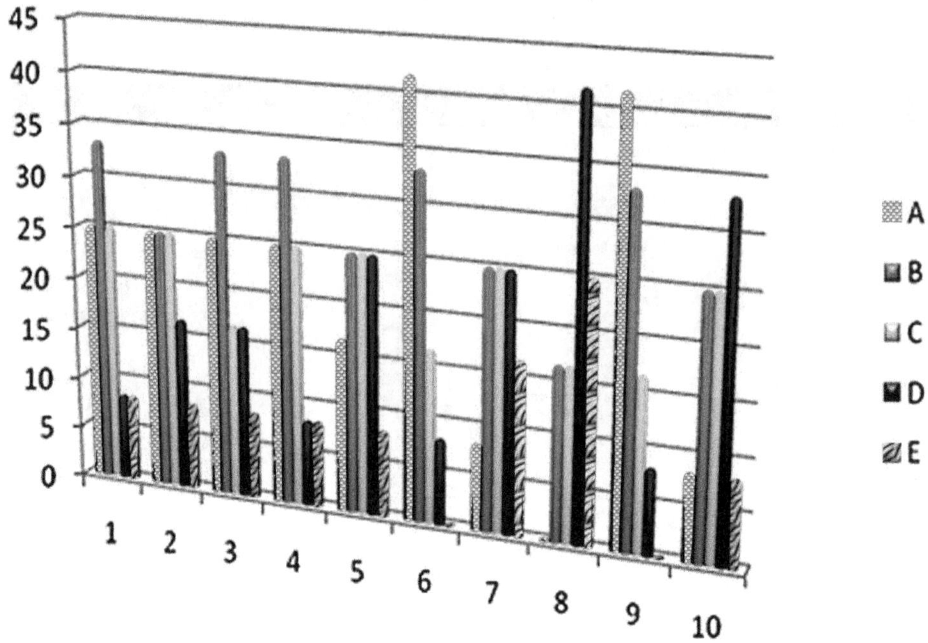

RECOMMENDATION FOR FUTURE RESEARCH

The conceptual design of the SECOnD portal is part of the overall continued research endeavor undertaken by South Africa's University of Zululand, Center of Excellence for Mobile e-Service, which aims at building an evolutionary system - GUISET infrastructure. It is conceptualized as a portal use case into which future e-services can be plugged. It would serve as a common point of entry for service integration and provisioning - a "bring-and-share" mode of utility computing; advertisement and deployment of available services and products, service categorization and cataloging. It would provide a template for customer specification of service parameter – service lookup and binding. The administration of membership (SMMEs) accounts subscription and billing aspect of the portal would also be addressed. Other portal functionalities also envisaged are: Business-to-Consumer portal capabilities - product or service information and ordering capabilities. Portlets are provided for specific product or service categories. These portlets are activated as needed by the portlet container (one portlet for each category of products or service information). Services exposed on the GUISET portal would have either a Native Component or Portlets as their backend. Thus, as the research work continues, it is intended that each of these requirements would be achieved afterwards as new research findings would be exploited to make the system fully realizable.

REFERENCES

Adigun, M. O. (2008). *GUISET: An Architecture and A Vision. Idina Hour Presentation at the Center of Excellence, e-Service*. KZN.

Akram, A., Dharmesh, C., Xiao, W., Yang, X., & Allan, R. (2005). A Software-Oriented Architecture for Portals Using Portlets. CCLRC e-Science Centre, CCLRC Daresbury Laboratory.

Alharbi, S., & Drew, S. (2014). Using the Technology Acceptance Model in Understanding Academics Behavioural Intention to Use Learning Management System. *International Journal of Advanced Computer Science and Applications*, *5*(1). doi:10.14569/IJACSA.2014.050120

Eilam, T., Appleby, K., Breh, J., Breiter, G., Daur, H., Fakhouri, S. A., & Wagner, H. et al. (2004). Using a Utility Computing Framework to Develop Utility Systems. *IBM Systems Journal*, *43*(1), 2004. doi:10.1147/sj.431.0097

Foster, I., & Kesselman, C. (1999). *The Grid: Blueprint for a New Computing Infrastructure*. Morgan-Kaufman.

Globus Toolkits. (2010). Retrieved from http://www.globus.org/toolkit/

Introduction to JSR-168. (n.d.). Last accessed on July 25, 2010 http://developers.sun.com/prodtech/portalserver/reference/techart/jsr168/

Jetspeed-2. (2010). Retrieved from http://portals.apache.org/jetspeed-2/

Liferay Portal Enterprise. (2010). Retrieved from http://www.liferay.com

Odusote, B., & Adigun, M. (2014). Technologically Enabling Resource-constrained Enterprises in Developing Nations Through The Implementation of E-Commerce on-Demand Service Portals for a Grid Utility Computing Platform. *Proc. 2014 IEEE 6th International Conference On Adaptive Science & Technology (ICAST)*. Retrieved from http://ieeexplore.ieee.org

Oscar, D., Salvador, T., & Sandy, P. (2007). Turning Portlets into Services: The Consumer Profile.*International World Wide Web Conference Committee (IW3C2)*.

Pagden, E. (2003). *The IT Utility Model-Part I, Sun Professional Services*. Sun BluePrintsTM Online.

Phillip, T. (2004). *Utility Computing - Identifying the Application Domain and Its Boundaries* (MSc. thesis). University of Zurich, Switzerland.

Rappa, M. (2004). The Utility Business Model and The Future Of Computing Services. *IBM Systems Journal*, *43*(1), 2004. doi:10.1147/sj.431.0032

Russell, M., Novotny, J., & Oliver, W. (2006). GridSpheres Grid Portlets. *Computational Methods in Science and Technology*, *12*(1), 89–97. doi:10.12921/cmst.2006.12.01.89-97

Web Services for Remote Portlets (WSRP). (n.d.). Retrieved from http://www.oasisopen.org/committees/tc_home.php?wg_abbrev=wsrphttp://www.oasis-open.org

Yang, X., Martin, T., Mark, H., Mark, C., Ligang, H., & Peter, M. (2006). Survey of Major Tools and Technologies for Grid-enabled Portal Development. *Proc. UK e-Science All Hands Meeting (NeSC 06).*

Yanli, C., Jian, C., Minglu, L., & Lei, C. (2006). Portlet-based Portal Design for Grid Systems. *Proc. 5th IEEE International Conference on Grid and Cooperative Computing Workshops (GCCW 05).*

ADDITIONAL READING

Díaz, O., Trujillo, S., & Pérez, S. (2007, May). Turning portlets into services: the consumer profile.*InProceedings of the 16th international conference on World Wide Web* (pp. 913-922). ACM. doi:10.1145/1242572.1242696

Migiro, S., & Adigun, M. (2005). ICTs, e-Commerce and rural development: The case of arts and crafts SMEs in rural KwaZulu-Natal. *Commonwealth Youth and Development*, *2*(3), 65–83.

Ning, J. Q. (1996). A Component-Based Software Development Model. *IEEE Software*, 390–391.

Sajjad, A., Jameel, H., Kalim, U., Lee, Y. K., & Lee, S. (2005). A Component-based Architecture for an Autonomic Middleware Enabling Mobile Access to Grid Infrastructure. *LNCS*, *3823*, 1225–1234.

Yang, X., Akram, A., & Allan, R. J. (2007). Developing portals/portlets using Enterprise JavaBeans for Grid users. *Concurrency and Computation*, *19*(12), 1633–1641. doi:10.1002/cpe.1200

KEY TERMS AND DEFINITIONS

Conceptual Design: This simply refers to what a design must entail or how it would work/function in order to realize the stated requirements.

SECOnD: Service-based Electronic Commerce On Demand paradigm is flexible and low-cost business model that enables customers to focus more on their core business by helping to satisfy their diverse business demands in a ubitquitous and timely manner.

Usability Evaluation: Usability Evaluation focuses on how well users can learn and use a product to achieve their goals. It also refers to how satisfied users are with that process.

Usability: Usability refers to the quality of a user's experience when interacting with products or systems, including websites, software, devices, or applications. Usability is about effectiveness, efficiency and the overall satisfaction of the user.

Utility Grid Computing: This is a service provisioning model in which a service provider makes computing resources and infrastructure management available to the customer as needed, and charges them for specific usage rather than a flat rate. Like other types of on-demand computing (such as grid computing), the utility model seeks to maximize the efficient use of resources and/or minimize associated costs.

About the Contributors

Charles Ayo holds a B.Sc, M.Sc, and Ph.D in Computer Science. He was the Vice Chancellor of Covenant University, Ota, between 2012 and 2016. His research interests include: Mobile computing, Internet programming, e-Business, e-Government, e-Health and Software Engineering. Prof. Ayo is a member of a number of international research bodies and has being an External Examiner to a number of Nigerian and Foreign universities. He has supervised about 200 postgraduate projects at Postgraduate Diploma, Masters and Ph.D levels, and has several publications in scholarly journals and conference proceedings. Professor Charles Ayo is a Senior Fulbright Research Scholar and an Awardee of Ford Foundation for Teaching and Innovations. He is well reputed Nationally and Internationally in the areas of application electronic and mobile technologies to governance, business, education and health among others.

* * *

Adewusi Adedeji is a young but vast scholar and a graduate student of the Department of Sociology, University of Ibadan. He has published in learned journals both locally and internationally.

Boluwaji A. Akinnuwesi is presently a Senior Lecturer (Full-Time) in Department of Computer Science in Lagos State University, Ojo, Lagos State, Nigeria. He obtained his B.Sc., M.Tech, and Ph.D. degrees, all in Computer Science and he focuses on Software Engineering and its Applications. His Ph.D. research focused on developing Neuro-Fuzzy based approach to evaluate the performance of distributed software system architecture. He was a Visiting Research Scholar at ICITD, Southern University, Baton Rouge, Louisiana, USA in 2010. He has several academic publications (i.e book, book chapter, journals and refereed conference proceedings). He is a Full Member of the following professional bodies: Computer Professionals Registration Council of Nigeria (MCPN), Association for Computing Machinery (MACM), Nigeria Computer Society (MNCS). He was awarded C.ITP (Chartered

Information Technology Practitioner) in 2009 by the Computer Professionals Registration Council of Nigeria. He has over 14 year of teaching and administrative experiences in the University. He had served as Acting Head of Department and also Acting Director of University Computer Center. He had also served in several University Management Committees. As Acting Director of University Computer Center, he initiated and implemented several ICT projects for the University. He is on the review board of a number of Computer Science journals/conferences. His research interest areas are: User-Centricity, Software Performance Engineering, Software Engineering, Database System, Expert System, Neural Networking, Fuzzy Logic, Artificial Intelligence, Computer programming concept, Technology Adoption.

Saleh Alwahaishi received his M.Sc. in Computer Science from New Mexico State University and his Ph.D. in Informatics from VSB-Technical University of Ostrava. Saleh Alwahaishi's research and development experience includes over 10 years in the Industry and Academia. Currently, he is working as a Management Information Systems lecturer at King Fahd University of Petroleum and Minerals. Before that, he worked as an Information Systems Instructor at United Arab Emirates University. He has authored/co-authored several refereed journal/conference papers and book chapters. His research interest included technology acceptance models, cultural issues in IT/IS adoption, website evaluation, e-learning technologies, and e-commerce.

Aregbesola Ayooluwa, an Information Scientist, is the Collections Development Librarian of Centre for Learning Resources, Landmark University, Omu-Aran, Nigeria. His research interest spans the field of Information Systems Adoption and Use.

E. O. Asani is a faculty member of the Department of Computer Science, Landmark University, Omuaran, Nigeria. He holds a Bachelor of Science degree in Computer Science from the University of Ilorin and Master's degree in Information Science from the University of Ibadan. He is currently pursuing his Ph.D in Cybersecurity with the Department of Computer Science, University of Ibadan. His research interest bothers on modelling of mathematical and statistical theories using soft computing tools to solve cybersecurity problems. He has attended and presented papers in both local and international conferences with his research efforts appearing in local and international peer reviewed journals. In August 2015, he was honoured at the iSTEAM MINTT conference held at the University of Benin for his contributions to scientific research. He is a motivator, teacher and leader, well loved by his students and mentees. Asani is goal oriented and student to great leaders, who believe in God and are committed to selfless service, integrity and hard work as an important hallmark of success.

Ambrose A. Azeta is a Senior Lecturer in the Department of Computer and Information Sciences, Covenant University, Ota, Ogun State, Nigeria. He holds B.Sc, M.Sc and Ph.D in Computer Science from University of Benin, University of Lagos and Covenant University respectively. His current research interests are in the following areas: Voice-enabled Systems, Technology-Enhanced Learning, Software Engineering, Internet Programming and Mobile Computing. He is a member of the Nigerian Computer Society (NCS) and Computer Professional Registration Council of Nigeria (CPN).

Melanie Baljko is an Associate Professor of Computer Science and Engineering at York University and the co-lead of the GaMay Lab. Her research concerns the design and evaluation of computational systems, such as those that provide assistance for individuals with disabilities and that are used clinically for speech disorders. She has a Ph.D. and M.Sc. in Computer Science from the University of Toronto and a B.Math from the University of Waterloo.

Richard Boateng is a technology researcher who focuses on developing, promoting and protecting ideas and concepts into sustainable projects of commercial value and development impact. Richard is an Associate Professor of Information Systems and the Head of the Department of Operations and Management Information systems at the University of Ghana Business School. He is also the Coordinator for the Building Stronger Universities Programme on E-learning and Problem-based Learning at the university. He is the associate editor of the Information Technologies & International Development Journal and also serves on the editorial board of the Information Development Journal. Richard's research experience covers e-learning, ICT for development, electronic governance, social media, electronic business, gender and ICT, mobile commerce, and ICT and health care at the national, industrial, organizational and community levels. He has consulted for UNDP, Ghana and UNESCO Accra Cluster Office on technology and communication development projects. Since 2006, Richard has published more than 30 articles in international journals, refereed conferences and book chapters.

Adefolarin Bolaj is a lecturer at the Department of Computer Science and Mathematics, Adeleke University Ede, Osun State, Nigeria. He had his BSc. in Computer Engineering and MSc. in Computer Science from the Department of Computer Science, Obafemi Awolowo Ile-Ife. He is at present a PhD candidate in the Department of Computer and Information Technology, Purdue University, USA. He has been into Web Programming and Information Security for about six years. He is a PHP-Zend Certified Engineer (ZCE), Cisco Certified Network Associate (CCNA), Microsoft Certified Professional (MCP) and Oracle Certified Associate

(OCA). He is also a Certified Ethical Hacker (CEH), Computer Hacking and Forensic Investigator (CHFI) and Ec-Council Certified Security Analyst. He is a member of the Institute of Electrical and Electronic Engineering (IEEE). His research focus is on Big Data Analytics, IoT and Cyber Security.

Emmanuel Eilu is a final year Information Technology PhD Student at the School of Computing and Informatics Technology –Makerere University. His areas of research and interest are in Human Computer Interaction (Usability and User Experience), Mobile User Experience, ICT for Development (Adoption of E-Government Systems; E-governance; E-voting; Biometric Voter Registration Systems; Electronic Civil Registers). He holds a master degree in Information Technology, specializing in E-government. Since 2007, Eilu has been teaching both post graduate and undergraduate programmes at the School of Computing and IT – Makerere University. He is also an adjunct lecturer at the International University of East Africa-Uganda. His major teaching area includes; Usability and User experience, E-government, E-governance, SAD, Information Systems Security, Information Systems Audit, Data Communication Networks and Research Methods. He has published over 15 research papers.

Foad Hamidi is a PhD candidate at the Lassonde School of Electrical Engineering and Computer Science at York University and a member of the GaMay Lab. His research interests include intercultural collaboration, assistive technology and digital activism. The underlying theme of his work is using technology for empowerment and community engagement.

Michaela Hynie is Associate Professor in the Department of Psychology, York University, Associate Director of the York Institute for Health Research (YIHR) and founder and Director of the YIHR's Program Evaluation Unit. Dr. Hynie is interested in engaged scholarship; working in partnership with students, communities and organizations, both locally and internationally, on research addressing complex social issues.

Meke Kapepo is a lecturer in information systems at the University of Cape Town. She lectures students on the academic development programme who have experienced gaps and disparities in their education. Her particular interest is in investigating the use of digital technologies to promote inclusive practices in higher education.

Kehbuma Langmia is an Associate Professor in the Department of Strategic, Legal and Management Communications and has recently been awarded a Fulbright

Scholarship to the University of Buea, Cameroon, Summer 2015. A graduate from the Mass Communication and Media Studies Program at Howard University in 2006, Dr. Langmia has extensive knowledge and expertise in Information Communication Technology (ICT) and Social Media. Since earning his PhD in Mass Communications and Media Studies from Howard University in 2006, he has written and published books, book chapters and research articles in prominent national and international journals. He is a recent recipient of a Fulbright Grant Award from the US Department of State to teach Masters and Ph.D. courses in the University of Buea, Cameroon this Summer 2015. He regularly gives keynote speeches on Information Communication Technology and Social Media in prominent national and international universities, including the Library of Congress, the National Intelligence University (Department of Defense, USA) and National Defense University (Department of Defense, USA); Madras Institute of Technology, India; ICT University, Cameroon and Covenant University, Lagos, Nigeria. He leads the Information Communication Technology Doctoral Consortium annually (under the ictforafrica.org conferences and seminars) through out Africa. His most recent book "Social Media: Pedagogy and Practice" which he co-edited and authored two chapters in it has been published by Rowman and Littlefield Publishers, 2014. This book has been adopted by some universities in the United States that have begun offering social media courses. The other, titled " Media Role in Changing Africa's Electoral Processes" still published by Rowman and Littlefield, 2014 has been adopted by some universities. Additionally, his most recent research articles include: "Cultural and rhetorical traditions in Africa/Black thinking" International Communication Gazette, 2015, 77 (5), 471-486, "Coping with smart phone distractions in a college classroom" co-authored with Amy Glass of Johns Hopkins University published in the current issue of the winter edition of Teaching Journalism and Mass Communication Journal, 2014, 4 (1), 13-23. The other is "Debunking the truth through a video documentary" to be published in the Fall 2014 of Journal of Third World Studies. Other publications include: "Moving the sign post? Resolving inter-village rivalry through a form of communication" in C. Onwumechili & I. Ndolo's book edition published by Rowman and Littlefield, Lexington books, 2013; "The secret weapon of globalization: China's activities in sub-Saharan Africa" in the Journal of Third World Studies, Vol. 28 (2), 2011 ; "Globalization and traditional cultures" in St. Armand, Kirk & O. Bolanle's edited book " Globalization and Digital Divide, published by Cambria Press, New York, 2011; "Media and Technology in Emerging African Democracies"with Cosmas Nwokeafor, published by the University Press of America, 2010.

Olumide Longe holds a PhD degree in Computer Science (Information Security Specialization) from the University of Benin, Benin City Nigeria; a Master's degree in Computer Technology (Information Systems) from the Federal University of

Technology Akure, Nigeria, a BSc in Computing from the University of Benin, Benin City, Nigeria and National Diploma in Electrical/Electronics Engineering from the Federal Polytechnic, Ado Ekiti, Nigeria. He is an alumnus MacArthur Scholar, an alumnus MIT Scholar and a Fulbright Scholar. Dr Longe has had a Progressive and accomplished academic and professional career demonstrating consistent success in research and development, professional impact, national and global recognition and assuring students success. He is Seasoned in conceiving management information systems theories (evidenced by scholastic publications), and initiatives in ICT diffusion and uptake. Longe has been named by Microsoft research as one of the leading experts on privacy and security in Africa. His current research focuses on Management Information Systems Security, social and enterprise informatics (using ubiquitous computing and social theories to aid management decision making processes). He is also involved in Cyber criminality profiling for identifying causation, assisting apprehension and providing treatments for cybercrime cases. He is the Convener of the Annual International Conference on Science, Technology, Education, Arts, Management and Social Sciences (iSTEAMS Research Nexus Conference and the iSTEAMS Honours Conference) and a Co-Convener of the Annual Africa Digital Week (www.africadigitalweek.com), an Adjunct PhD Supervisor in Computing and Information Security at the Accra Institute of Technology, Accra Ghana and the Co-Founder of the Creative Research and Technology Evaluation Network (CREATE Network), Dr. Longe is a Fellow of the African Scientific Institute. He is a Professor of Computer & Information Security and Dean, Caleb Business School, Imota, Lagos State, Nigeria.

Tabisa Mayisela is a lecturer and digital literacy coordinator at the University of Cape Town. She teaches in the Post graduate diploma in educational technologies. Her research interests are: Pedagogy and technology integration into the curriculum; and Approaches to integrating digital literacies into course curricula.

Patrick Mbullo has a MA in Anthropology from York University. He is currently at Maseno University in Kisumu, Kenya.

Susan McGrath is Professor, School of Social Work and Centre for Refugee Studies, York University.

Francis Nchuchuwe (Ph.D) was born in Lagos on December 4, 1959, Dr. Nchuchuwe hails from Obikwele in Ndokwa-East Local Government Area of Delta State. He attended the Local Authority Primary school, Ikeja between 1966-1972 and later proceeded to Victory High School, Abule Onigbagbo, Mary-Land, Ikeja between 1972-1977 where he obtained the West African School Certificate

(WASC). He bagged a Bachelor of Science degree (Double Honours) in Political Science and Public Administration with a first class in 1997, a Master of Science degree in Public Administration in 1998 and Doctor of Philosophy (PhD) degree in Public Administration, with specialization in Local Government in 2006, all from the prestigious great University of Benin, Benin City, Nigeria. (Igboamalu Nwafor - Contributing Author) Mr. Igboamalu, Peter Nwafor teaches Computer Studies with the Enugu State Secondary School Management Board. He holds a B.Sc. (Hons.) and Master Degrees in Computer Education from the Department of Vocational Teacher Education, University of Nigeria Nsukka. He is currently a Doctoral student in the Department of Computer and Robotics Education, University of Nigeria, Nsukka with special interest in teaching and learning research initiatives (TLRI) aimed at improving academic achievement of learning outcomes. Mr. Nwafor hails from Umulokpa in Uzo-Uwani LGA of Enugu State. He is happily married with two children.

Babafemi O. Odusote is a 1st class Computer Science graduate of Covenant University. He holds a M.Sc. with Distinction in Computer Science in 2011. He is currently on his Ph.D Program in the same Institution. His research interests include: Software Engineering, Mobile Computing, Grid Computing, Service-Oriented Computing, e-Learning, e-Commerce.

Olusegun Ogundiwin is a graduate of Computer Science. He is currently on his Master's degree programme in Computer Science. His research interest areas are: Software System development, Management Information System, System Analysis and Design.

Kemi Ogunsola obtained Master and Ph.D. degrees in Information Science, from Africa Regional Centre for Information Science, University of Ibadan, Nigeria, where she also lectures. Her research focuses on applying Information and Communication Technologies (ICT) to development, particularly as it relates to governance. She is a recipient of Next Generation Social Science Research in Africa fellowship, awarded by the Social Science Research Council (SSRC), New York, USA. She is a fellow of Carnegie Corporation of U.S.A. She has published articles in both local and international journals. She has also made presentations at international conferences.

Foluke Okocha is an information professional with research interests in e-commerce and e-governance.

Chijioke Jonathan Olelewe is currently a Lecturer in the Department of Computer & Robotics Education, University of Nigeria Nsukka. Dr. Olelewe, holds a B.Sc.

(Hons.) and Master Degree and Ph.D in Computer Education from the Department of Vocational Teacher Education, University of Nigeria Nsukka. He has special interest in Teaching and Learning Research Initiatives (TLRI) aimed at improving students' achievements in both psychomotor and affective domains.

Aderonke A. Oni holds Ph.D in MIS. Her research interests are in: e-commerce, e-government, and technology adoption. She has published over twenty-five articles in reputable international peer-reviewed journals, proceedings, and chapters in books. She is a lecturer in the Department of Computer and Information Sciences, Covenant University, Ota, Nigeria. She was a post-doctoral research fellow at International Center for Information and Communication Technology, Southern University, Baton Rouge, USA between 2014 and 2015.

Meke Shivute-Kapepo is a Lecturer of Information Systems 1 (ADP) in the department of Information Systems at the University of Cape Town. She is interested in Information and Communication Technologies for development in education. More specifically her work examines pedagogy, curriculum, online learning and teaching Information systems in higher education.

Václav Snášel is a researcher and university professor. He is also a dean at the VSB-Technical University of Ostrava. His research and development experience includes over 25 years in the Industry and Academia. He works in a multi-disciplinary environment involving artificial intelligence, multidimensional data indexing, conceptual lattice, information retrieval, semantic web, knowledge management, data compression, machine intelligence, neural network, web intelligence, data mining and applied to various real world problems.

Aida Tetegan is a graduate of Computer Science. She is currently on her Master's degree programme in Computer Science. Her research interest areas are: Management Information System, System Analysis and Design.

Index

T

U

V

W

www.ingramcontent.com/pod-product-compliance
Lightning Source LLC
LaVergne TN
LVHW062120200926
848484LV00007B/1721